MW00845305

U.S.S. Albacore
Forerunner of the Future

The Albacore, *AGSS-569 in drydock at night.*

U.S.S. Albacore
Forerunner of the Future

by

Robert P. Largess

and

James L. Madelblatt

The Portsmouth Marine Society

Publication Twenty-Five

Published for the Society by

Peter E. Randall

PUBLISHER

Printed in the United States of America

Designed and Produced by
Peter E. Randall Publisher
Box 4725, Portsmouth, NH 03801

A publication of
The Portsmouth Marine Society
Box 147, Portsmouth, NH 03801

Library of Congress Cataloging-in-Publication data
Largess, Robert S. 1945-
U.S. S. Albacore : forerunner of the future / by Robert S. Largess and James L. Mandelblatt
p. cm. -- (Publication / The Portsmouth Marine Society ; 25)
Includes bibliographical references and index.
ISBN 0915819-24-4 (alk. paper) -- ISBN 0-915819-25-2 (pbk. : alk. paper)
1. Albacore (submarine) I. Mandelblatt, James L. II. Title. III. Publication (Portsmouth Marine Society) 25.
VA65.A25 L37 1999
359.9'3832'0973--dc21

99-047504

The Portsmouth Marine Society is grateful to the following *Albacore* Officers and Crew Members, and Sponsors whose support has helped to make this publication possible.

OFFICERS AND CREW

Norm Bower (COB)
Keith E. Perry
LCDR Ronald D. Hines
Edward L. Crowley-ETSA-1954
Henry "Shaky" Graves
John J. Sgro ET3 SS
David R. "Pete" Peterson IC2 (SS)
LCDR (Later Vice Adm.) Lando W. Zech, Jr., 3rd Commanding Officer, USS *Albacore*
Keith W. Kraft, Plank Owner 1953-54-55
Francis P. Lyons
Jack E. Hunter (LCDR,USN)
Herbert H. Zahn MMe/SS Ret. Crew member 1962-1963
Joseph E. Dinelli
Theodore F. Davis
Jack R. Young
Geo. C. "Buzz" Sawyer Jr. SO3 SS
Russell Schondorf MM2, May 1966-October 1968
Harry T. "Bud" Fisher
Kenneth Latchaw PM3 SS B184930
John Gandiello SN (SS) (1971-1972)
Roger Romito (SNSS) Plank Owner
LCDR John McCarthy, USN (RET)
Wallace A. Greene, Captain USN (Retired) C.O. USS *Albacore*, 8/24/60-1/11/62
LCDR Bernarr M. Bowdoin USN (Ret)
Lt. William A. Miller, USN
August 1955-August 1957
Barry V. McCutchan
Patrick J. McGroarty
Dennis L. Hayes
Robert P. Freytag
Capt. Michael J. Donahue JAGC USNR-RET
Fred W. Saunders
Alex T. Szymanski
Dave Cahill MMC (SS) 1963 to 1969
LCDR Jon L. Boyes, USN (Ret)
Allan H. Huhtala
Paul C. Azarian
Stewart E. White
Theodore James Karmeris Jr.
Capt. Roger H. Kattmann (Deceased) Commanding Officer-USS *Albacore* Aug 1967 to July 1969 (Given by Mrs. Roger H. Kattmann)
Russell V. Davis Jr. ET2 (SS)
Ronald F. Poloske
James E. Kraut
Stan Zajechowski

SPONSORS

Mark G. Phillips
Warren and Dorothy Pratt
Vice Admiral Jon L. Boyes, USN
William W. Seaward Jr.
Ricci Lumber
Richard E. Sullivan
Ferris G. Bavicchi
Ambit Engineering
Prof. and Mrs. Eugene Allmendinger
Robert A. Lincoln
Daniel H. Vickery
Bruce R. Graves
Susan Sawtelle
Squalus Memorial Chapter, U.S. SUB VETS WW II
Russ and Barbara Van Billiard
Peter Pierce Rice
William W. Howells
Muriel G. S. Howells
CWO Mark L. Anthony (Ret)
Eileen Foley
Roland L. Couture
Joseph and Jean Sawtelle
Cyrus B. Sweet III
Dan Franzoso
John T. Robinson, PE

Other Portsmouth Marine Society Publications:

1. *John Haley Bellamy, Carver of Eagles*
2. *The Prescott Story*
3. *The Piscataqua Gundalow, Workhorse for a Tidal Basin Empire*
4. *The Checkered Career of Tobias Lear*
5. *Clippers of the Port of Portsmouth and the Men Who Built Them*
6. *Portsmouth-Built, Submarines of the Portsmouth Naval Shipyard*
7. *Atlantic Heights, A World War I Shipbuilders' Community*
8. *There Are No Victors Here, A Local Perspective on the Treaty of Portsmouth*
9. *The Diary of the Portsmouth, Kittery and York Electric Railroad*
10. *Port of Portsmouth Ships and the Cotton Trade*
11. *Port of Dover, Two Centuries of Shipping on the Cochecho*
12. *Wealth and Honour, Portsmouth During the Golden Age of Privateering, 1775-1815*
13. *The Sarah Mildred Long Bridge, A History of the Maine-New Hampshire Interstate Bridge from Portsmouth, New Hampshire, to Kittery, Maine*
14. *The Isles of Shoals, A Visual History*
15. *Tall Ships of the Piscataqua: 1830-1877*
16. *George Washington in New Hampshire*
17. *Portsmouth and the Montgomery*
18. *Tugboats on the Piscataqua: A Brief History of Towing on One of America's Toughest Rivers*
19. *Rambles About Portsmouth*
20. *John Paul Jones and the Ranger*
21. *Constructing Munitions of War: The Portsmouth Navy Yard Confronts the Confederacy, 1861-1865*
22. *In Female Worth and Elegance: Sampler and Needlework Students and Teachers in Portsmouth, New Hampshire, 1741-1840*
23. *Gosport Remembered: The Last Village at the Isles of Shoals*
24. *Heavy Weather and Hard Luck: Portsmouth Goes Whaling*

Contents

To my parents
Zoe McCombs Largess
and
Cmdr. George J. Largess, USNA '39
My first and most lasting inspiration

Acknowledgments

THIS BOOK COULD NOT HAVE BEEN COMPLETED without the generous help of many, many people. First and foremost, I would like to thank Harvey S. Horwitz, my collaborator on several earlier articles. We first discussed this project about 1988. Harvey did a monumental amount of work on it, interviews, photography, and correspondence during the early years. However, a new career as superintendent of schools and then serious health problems left him simply unable to continue after 1993.

Next, thanks to Jim Mandelblatt, historian of the *USS Requin* who came aboard in the fall of 1996. Without him, this book would never have been finished.

And thanks to Russ Van Billiard, Peter Randall, and Joe Sawtelle, representing the Portsmouth Submarine Memorial and the Portsmouth Marine Society, who have waited so long, more or less patiently, for this book to be completed.

I can only say two things in mitigation. First, I have been fighting my own "Cold War" during these years in the public schools of an American inner city. Indeed, tracing the story of *Albacore* has been a relief and distraction from daily frustration and failure: a story of intelligence, determination, and courage actually succeeding, proving that sometimes, things can work, and hard effort be rewarded.

But a second reason is that in the early years much to do with the Navy's submarine program was still classified and much archival material was simply closed to me. The end of the Cold War changed all that and about 1994 the walls began to come down. Also, very little was published on *Albacore* or the sub program before then. When I began, all I had to work with were short articles by David Merriman, Ted Davis, Jon Boyes, and Gene Allmendinger and Harry Jackson. In effect,

effect, I had to learn everything from scratch; it looked like I was going to have to write the entire history of the submarine program to understand the significance of anything *Albacore* did. Originally, I expected this book would be highly scholarly and thoroughly footnoted; but at the rate I was going this would have taken thirty years.

However, very fortunately for me two superb books appeared which told the story of the postwar submarine program and between them provided a thorough context to the story of the *Albacore*. These are Gary Weir's 1993 *Forged in War; the Naval Industrial Complex and American Submarine Construction, 1940- 1961*, and Norman Friedman's *U.S. Submarines since 1945; an Illustrated Design History*. Dr. Friedman concentrates on technology and tactical concepts, while Dr. Weir focuses on scientific, industrial, and Navy policy and cooperation; each very effectively complements the other. The reader will find both works cited repeatedly but they have informed this book so deeply I can hardly cite them enough.

Thanks also to Dr. Weir for his personal assistance and advice, and thanks to the archivists he directed me to: Carla Bowen at the National Academy of Sciences, Dixie Gordon and Sid Reed at the NAS' Naval Studies Board, Mike Walker at the Naval Historical Center, Wes Pryce at Navy Labs Archives, and those at the National Archives. Thanks also to the photo archivists at the Naval Historical Center and the Naval Institute.

Above all, my deepest thanks to the officers and men of *Albacore*, and the many others, family members, scientists and engineers, shipyard workers, volunteers of the Portsmouth Submarine Memorial, who granted interviews, answered questions, provided rare and vital documents, photos, and memorabilia. Too many to mention here, their names and voices will be met with in the text or photo captions. But one exception must be Capt. Ted Davis, whose article "Will Our Subs Have A Fighting Chance?" was the original inspiration for the book and who was first to help. Much of Chapter 4 is in his own inimitable voice.

I must also apologize to the many individuals I didn't contact. I have many, many of your names given to me by your old friends and associates. But the book had to be finished: I had to content myself with talking to representative members of each of the significant classes of people who formed part of the story of *Albacore*.

Finally, one might ask what this book adds to Dr. Weir and Dr. Friedman's work, besides giving the story of *Albacore* herself in full detail. I have been concerned throughout to give the human, experiential side of the submarine story. Who were the people, what did they do, what did it mean to them, and what did it feel like to be there?

As Bob Johnston, longtime Portsmouth Naval Shipyard employee and active in the effort to save *Albacore* and bring her back to Portsmouth, put it:

> She had been put into mothballs just as she sat—notes on the notepads, manuals open, tools in use left where they were dropped, like the person whose station it was had just left for a moment and would be back in time to finish the cigarette in the ashtray. It's a strange thing, how a ship is alive, a living organism, so many complicated things going on when the crew is aboard. All the human feelings blend in to create a mind or spirit of her own. Then they walk off, and she's just iron.

Albacore was people, not just iron. (And, of course, before she was iron she was ideas in the minds of her creators and designers.) I have been at pains to let these people speak in their own voices as much as possible. My goal is to, as the Bard puts it, "on your imaginary forces work," and to make the living organism that was *Albacore* live again in your mind.

And last of all thanks to my wife Jeannine who has helped and supported me on this project in a thousand ways since its inception. Thanks to my children who have had to share my thoughts and fondness with this piece of iron for all these years. And thanks to my first inspiration, my father CMDR George J. Largess, Annapolis Class of 1939. It was early apparent that I would be a bookworm, not a sailor (and could this be inherited from my mother and her love for Shakespeare and John Donne, perhaps), but if you should guess that these pages reveal a vicarious love and passion to the point of madness for ships and the sea, well, you just might be on to something.

Robert P. Largess
Boston, Nov. 8, 1998

USS Albacore, *(AGSS-569), as she looks today as the Portsmouth Submarine Memorial. The first truly streamlined submarine, pioneer of many technological advances incorporated in the US nuclear submarine fleet, historian Gary Weir has called her "the most significant submarine ever built." (Peter Randall)*

Introduction

USS Albacore*: Portsmouth Submarine Memorial*

MOTORISTS PASSING THROUGH PORTSMOUTH, NEW HAMPSHIRE, on Route I-95 on their way to Maine will notice a sign for the USS *Albacore* at exit 7, the Market Street Extension, just before crossing the high-level bridge over the Piscataqua River. Portsmouth harbor is *Albacore*'s birthplace, as well as her final home. Portsmouth's maritime heritage, its ties to the U.S. Navy and to the Submarine Service in particular, runs deep. Britain built the 54-gun frigate *Falkland* here in 1690, the first of many warships. An island across the river in Kittery, Maine, was acquired by the U.S. government in 1800, to become the Portsmouth Naval Shipyard (PNS).

PNS built its first wooden warship for the Navy in 1815. The yard launched its first submarine in 1917. The *L-8* was the first of 134 submarines to be built here, including 80 of the famous "fleet boats" during World War II, and 10 nuclear submarines between 1958 and 1971. PNS became a major center for the development of submarine technology, especially in the revolutionary decades after World War II, which saw the birth of the nuclear submarine—and of *Albacore*.

Many of the shipyard workers who built these submarines were from Portsmouth and Kittery; many of the men and their sons served in the Navy's Submarine Service. And many Navy men, brought to Portsmouth by the Submarine Service, have chosen to retire and make their homes here. Thus, the Portsmouth Submarine Memorial holds deep meaning for both the region and submariners alike. However, the Memorial's centerpiece is not one of the seven Portsmouth-built fleet submarines of World War II preserved today. (They are the *Lionfish, Drum, Batfish, Bowfin, Pampanito, Torsk,* and *Requin*. The post–WWII

This bow-on view shows Albacore's *"body of revolution" hull with its circular cross section like a rocket, torpedo—or blimp. To the greatest possible extent anything creating drag or unbalanced hydrodynamic forces was eliminated from* Albacore's *"teardrop" hull, to simplify her primary mission, the study of submarine maneuvering and control at high submerged speeds. (Peter E. Randall)*

Regulus-missile launching *Growler* makes a ninth Portsmouth built.) It is the experimental submarine *Albacore* (AGSS-569). She was laid down at PNS on March 15, 1952, launched on August 1, 1953, commissioned on December 5, 1953, and began operations as a test vehicle on February 5, 1954. Decommissioned on September 1, 1972, *Albacore* joined the Navy's mothball fleet at Philadelphia and stayed there until towed home to Portsmouth, arriving there on April 29, 1984, to become a memorial and museum. She neither served in war nor took part in any military action; throughout her career, *Albacore* never carried any armament.

With her tubby, blimplike "teardrop" or "body of revolution" hull, *Albacore* was the first wholly streamlined submarine. She twice claimed the record for the fastest submarine in the world—that is, highest submerged speeds. Indeed, *Albacore*'s performance was phenomenal: As built, she made 27 knots at a time when no other submarine could do much better than 18 knots submerged. Modified in the 1960s, *Albacore* could make speeds submerged similar to the top speeds of destroyers—

around 35 knots—better than any contemporary nuclear submarine. She contributed not only her hull form and speed to most of the U.S. nuclear submarine fleet, beginning with the *Skipjack* class of 1958 (which temporarily took away *Albacore*'s speed record), but many other essential elements of her technology as well. Indeed, she tested a variety of advanced ideas that were never adopted in service.

As a ship memorial, *Albacore* is unusual in that she is preserved on dry land, mounted on huge concrete cradles in an excavated, permanent "dry dock." There, her enormous whale-shaped hull is completely and dramatically visible. Her smoothly curved black hull, with its impressive fins, evokes the image of some great sea creature, rather than a ship. And most appropriately so, for *Albacore* was the first submarine truly designed to use the waters of the sea not merely as a cloak of invisibility, but as her primary medium of operation. From *Albacore* came the ability of today's submarines to maneuver like aircraft in a three-dimensional medium, a water envelope much more dangerous and less well known even today than that of the air.

The design of her hull was worked out by the scientists of the Navy's David Taylor Model Basin (DTMB), its huge research laboratory dedicated in 1939, in Carderock, Maryland. In turn, she provided these scientists, through an extensive testing program carried out over many years, with the basic data on the hydrodynamic forces that affect submarines at high-speeds and in extreme turns and dives, establishing the "flight profiles" that enabled the behavior of the Navy's nuclear submarines to be predicted and controlled under all conditions.

The *Albacore* represents the remarkable partnership between the U.S. Navy and the American scientific community that developed during World War II and came to fruition in the decade that followed. To some extent, a happy combination of circumstances helped ensure that her discoveries had an impact. The fact that she was unarmed meant that the Navy was content to entrust her completely to the control of scientists; and the fact that she was a full-scale, active-service, Navy-manned submarine ensured that the Navy took her achievements seriously—accepted them as "for real." Dr. Gary Weir, in his *Forged in War,* his scholarly study of the creation of the Navy's submarine force, says:

"AGSS 569 became the vehicle through which revolutionary ideas generated by the close relationship between the components of the naval-industrial complex found their way to the operational forces. *Albacore* embodied the essence of the future of undersea warfare . . ., changed forever the character of the world's submarines, and is, in a comprehensive sense, the most significant submarine ever built."

But how was the *Albacore* first conceived and designed? What did

Portsmouth Naval Shipyard, birthplace of Albacore *and her main base of operations for her 19-year career. PNS built 134 Navy submarines between 1917 and 1971, including 10 nuclear boats.* Albacore *stands as a memorial not only to the Navy's subs and submariners but also the skill and ingenuity of PNS' engineers and shipyard workers, and to Portsmouth, New Hampshire, and Kittery, Maine, with their deep ties to the submarine service. (PNS Museum)*

she do and achieve in her career of nearly two decades? Who were the people who created her, served aboard her, and made her a success?

This is her story, but also the story of the people who made her live: the scientists who conceived her, the engineers and naval architects who evolved their concepts into the design of a real ship, the PNS workers who constructed her, the officers and Navy men who took her to sea and made her do what no other submarine had done before. She was, as her motto says, *Praenuntius futuri*—forerunner of the future.

I *Why the* Albacore?

ON MAY 2, 1982, THE BRITISH NUCLEAR ATTACK SUBMARINE *Conqueror* torpedoed and sank the Argentine cruiser *General Belgrano* and in effect decided at a stroke the Falklands War. The substantial Argentine fleet had no alternative but to return to port and stay there, absolutely defenseless against the three British nuclear submarines operating in the area. This incident, the only wartime action by a nuclear submarine, illustrates the simple fact that the submarine has become the decisive weapon in naval combat at sea. In the last decade of the Cold War, the U.S. "maritime strategy" depended on fast and powerful *Los Angeles*–class nuclear attack submarines to screen the carrier task forces and precede them, "sanitizing" their areas of operation from the threat of Soviet *Victor*- and *Charlie*–class nuclear attack submarines. Nuclear submarines have become the seaborne equivalent of fighter aircraft, the primary weapon for ship-to-ship combat.

Even more significant was the development of the nuclear ballistic missile submarine, beginning with the *George Washington* in 1959. The ultimate purpose of nuclear weapons is to deter the use of nuclear weapons. Yet as long as both sides depended on land-based ballistic missiles and bombers, the possibility existed that one side would be tempted to strike first, seeking to wipe out the other side's nuclear capacity to strike by surprise. Both the United States and the Soviet Union had to be ready to launch within minutes, creating the possibility of accidental nuclear war. (President Kennedy and Chairman Khrushchev established the so-called Hot Line to prevent this.) But

The USS Holland, *SS-1, the US Navy's and the world's first successful submarine. Many see her as the forerunner of the* Albacore *with her tubby, fish-like, reasonably streamlined hull. But she was shortranged and slow, and very quickly it appeared that submarine would require better surface speed and seaworthiness for strategic and tactical mobility. (NHC)*

the *Polaris* submarines could simply vanish into the deeps. None were ever located or tracked by the Soviets during the Cold War. Because they could not be found, these submarines could not be attacked, eliminating the possibility of a preemptive first strike. Slowly, the nuclear holocaust seemed less and less a real possibility. Today, the Cold War is over; perhaps the nuclear "genie" is back in the bottle, and mankind really can breathe easier.

Albacore, the "forerunner of the future," was one of the cornerstones of this achievement, along with the *Nautilus* and nuclear power in 1955. But to understand *Albacore*'s place in the history of submarines, it is necessary to understand the submarine's place in history.

It is worth noting that there have been only three potentially decisive submarine campaigns in history. Two of these were the German U-boat campaigns of World War I and World War II. Both brought Germany close to a victory over Great Britain and, from the viewpoints of the Royal Navy and U.S. Navy, represent *anti*submarine campaigns. The third is the successful American submarine campaign conducted against Japan during World War II.

In World War I, no one had more than a handful of primitive submarines at first, nor any clear idea of how to use them. But in the deadlock created by trench warfare on the western front, the submarine burst like a bomb. Germany discovered that its U-boats could range with impunity throughout British waters, sinking at will the merchant shipping that teemed there. The British countered with huge numbers of small warships, aircraft, and blimps. They frequently encountered U-boats and forced the German submarines under, but failed hopelessly at sinking them or protecting merchant shipping. The British simply could not cover enough ocean to find more than the occasional U-boat, and they had absolutely no means of detecting the submerged U-boat.

Germany declared unrestricted submarine warfare for the first time in February 1915, but was forced to back off by intense American pressure after the sinking of the *Lusitania*. In January 1917, however, a desperate Germany decided to accept war with America as the price for a quick victory over Britain and again declared unrestricted submarine warfare. The number of sinkings soared; Britain was down to six weeks of essential supplies before the tide turned. What saved Britain was the decision to concentrate merchant shipping into escorted convoys, screened by aircraft and destroyers.

The submarine was the first and ultimate weapon of "stealth"; submergence gave it the invulnerability of invisibility. Its weakness was that it traded its mobility for this invisibility. Until 1945, its absolute maximum submerged speed was 10 knots, sustainable for an hour at best before exhausting its batteries. At 3 or 4 knots, a submarine's battery charge could last about 24 hours, about the same as the oxygen inside the submarine. Thus, when a submarine was forced to use its surface speed to overtake a convoy in order to gain an attacking position on its bow, it could be spotted by the surface escorts and forced under. Sunk or not, the submerged submarine would be unable to overtake the convoy.

In World War II, Adm. Karl Doenitz conceived the *Rudeltaktik*,

Conventional U-boats such as the two on the left posed the greatest threat to Britain's survival in 1940 and 1941. They attacked convoys in "wolfpacks" at night and on the surface; they were defeated by radar which robbed them of their surface mobility and invisibility. The German solution was the Type XXI U-boat (here U-3008) with high submerged speed and endurance. (NHC)

or "wolf pack" as a means of defeating convoys by attacking on the surface with groups of U-boats at night. With their low silhouettes, the U-boats were nearly invisible in darkness and their 17-knot surface speeds were better than those of many escorts. Inside the convoy, they attacked the merchantmen with torpedoes on the surface, with impunity at first.

Meanwhile, the British had gone from the simple hydrophones of World War I, able to detect the sounds of a submerged submarine, to active sonar, or Asdic, in World War II, which used a sound impulse whose echo revealed the submarine's range and bearing. It was short ranged, usable only at low speeds, and unable to indicate depth. But it *could* hold a contact. Submerging could be a fatal mistake. The submerged submarine‘s only hope to escape was by running slow and silent, changing depth, and creeping away when the surface escort ran in with depth charges.

The U-boats of World War II were ultimately defeated by a gigantic effort, a huge force of ships and aircraft, together with many scientific advances. What threw the U-boats back to square one however, was the invention of radar, which robbed the U-boats of their invisibility and the use of the surface at night, defeating Doenitz's *Rudeltaktik*. By May 1943, German submarines were suffering ruinous losses. German science then sought its own countermeasures. The *schnorchel* breathing device enabled the U-boat to use its diesels submerged, freeing it from the need to surface. The *schnorchel* (or snorkel) gave the U-boats some chance of survival, but they achieved little.

The real solution was higher submerged speed. Even an improvement of a few knots would make it impossible for a surface vessel to hold the submerged contact with its sonar. A submarine fast enough to overtake a convoy submerged (10 to 12 knots) would be almost immune to counterattack. Doenitz explored two approaches: an advanced means of propulsion making the submarine independent of the atmosphere, and streamlining to decrease the drag of the submerged submarine.

The former, the hydrogen peroxide "Walter" turbine, was unsuccessful; independence from the atmosphere would become a reality only with the advent of nuclear power. The second approach produced the Type XXI U-boat, one of the weapons, like the jet aircraft and the V-2 rocket, that the Germans hoped would change the course of the war. When it fell into the hands of the Allies at the end of the war, they were astonished by the Type XXI's performance. The Type XXI was the first operational streamlined submarine, not in the total sense of the *Albacore* with her teardrop hull, but in the partial sense of eliminating all projections and angular lines, sources of underwater drag and turbulence, from the typical long, narrow, cigar-shaped hull. With increased power and battery capacity, this gave the Type XXI a submerged endurance of 18 knots for 1½ hours, or 14 knots for 10 hours, or 5 knots for 60 hours. This is compared to a conventional U-boat, which

The Type XXI achieved its high submerged speed by increased battery capacity and electric motor power, plus drastically decreased resistance achieved by streamlining - not in the sense of Albacore's *teardrop hull, but by removing projections and angular lines from a conventional cigar-shaped submarine hull. (NHC)*

could do 6 knots for three-quarters of an hour, or $1\frac{1}{2}$ knots for 30 hours. (The *Albacore* in her original configuration could do 27.4 knots for 30 minutes, 21.5 knots for 1 hour, and at low speeds had an estimated 75 hours of submerged endurance.)

Determined to master its technology, the United States, Great Britain, and Russia tested Type XXI U-boats captured as war prizes. For four years the U.S. Navy operated the U-2513 and U-3008 at

The USS Tang *(SS-563), the US Navy's version of the XXI formula, conventional-hulled but streamlined. She embodied many innovations and shared much new technology, including her radial "pancake" diesels, with the* Albacore. *At the end of World War II, the Type XXI was beyond the capacity of any counter-measures; it was essential for the US to master its technology and learn its secrets. (US Navy)*

Portsmouth. It became apparent immediately that the construction of substantial numbers of Type XXI clones by the Soviet Union would pose an insuperable danger, threatening a third, almost hopeless Battle of the Atlantic. The beginning of the gigantic effort of 47 years to counter this threat was the creation of similar submarines of advanced performance to serve first and foremost as lowly target vessels to train our antisubmarine warfare (ASW) forces. Wisely, it turns

The US Navy's chief nightmare in 1945 was the creation of a large fleet of Soviet Type XXI clones, threatening a new, unwinnable Battle of the Atlantic. This did materialize in the form of 236 Whiskey-class *boats. However, by the time they and many, many even better boats actually appeared, the US Navy had made strenuous efforts to develop the technology to defeat and surpass them - including the* Albacore. *(NHC)*

out, for the Soviet Union had long seen the submarine as the key to sea power. Determined not to make the Germans' error of starting a war with a tiny submarine force, they built 236 *Whiskey*-class Type XXI copies, added to their huge World War II fleet of more than 300 conventional submarines.

Different lines of thought suggested several solutions. One was to streamline the World War II fleet boats; this led to the *Guppies,* which formed the mainstay of the Navy's submarine force through the 1950s. The second solution was to build our own Type XXI clone, conventionally hulled diesel boats optimized for submerged speed and performance, embodying many advanced features; this became the *Tang* class. But some voices, notably that of Kenneth S.M. Davidson of the Stevens Institute of Technology, argued that juggling existing designs would yield a second-rate answer. The real solution would be a totally unconventional approach—the *Albacore*.

Some years later, Professor A.G. Hill of M.I.T. said, "... I think all of us started this job with the distinct feeling that this country was very much like a man with an inoperable cancer; that he might linger a while but that he was really done. We changed this opinion before finishing but the change was gradual." What was not apparent was that the submarine itself would be a large part of the solution.

Albacore's Heritage— U.S. Submarines in World War II

The submarine service, which grappled with the innovations and challenges of the Type XXI U-boats, had just won history's third and sole successful submarine campaign, sinking 1,314 Japanese ships, totaling more than 5.3 million tons, or 55 percent of all Japanese shipping losses during World War II. More than any other single factor, U.S. submarines were responsible for Japan's economic and logistic collapse. They paid heavily for this success, however, losing 52 submarines and 3,502 men, a casualty rate of 22 percent, highest for any branch of the U.S. military during the war. Strangely enough, the first two years of this campaign were a heartbreaking debacle, due to lack of readiness, faulty tactical doctrine, faulty peacetime training, and faulty weapons. Above all, our boats were plagued by ineffective torpedoes, which tended to run beneath their targets and used a magnetic exploder that did not work. Sadly, no live firing tests had been conducted during peacetime using these torpedoes.

Finally, 21 months after Pearl Harbor, tests were carried out that identified the problems with the American torpedoes. Given an effective weapon, U.S. submarines began to press home their attacks, sinking Japanese warships and merchant ships, providing distant reconnaissance of Japanese fleet operations, and rescuing downed American and Allied pilots (such as president-to-be George Bush). Submarines landed Marine raiding parties, bombarded enemy shore installations, and conducted photoreconnaissance missions of landing sites. Out of this experience—this cauldron—came the Navy's postwar submarine service. Its officers provided the skippers for the *Guppies,* the *Tangs,* and the early nuclear boats. Members of the Pacific command staffs found themselves in Washington, wrestling with the role of the submarine in the future. Two admirals in particular, Charles B. Momsen and Charles W. Styer, had played major roles in overcoming the flaws in the torpedoes and tactics that crucified American submarines in the

While the U-boats were losing a desperately-fought Battle of the Atlantic, US submarines were winning the world's only successful, decisive submarine campaign, against Japan. Here is the brand-new "first" Albacore, *SS-218, at Groton on May 9, 1942. Her high enclosed bridge and provision for little surface armament reflect the Navy's tactical concepts early in World War II. US subs operated submerged in daylight to avoid planes, sacrificing much mobility. They planned to track targets by sonar and fire torpedoes at great depths; things wholly beyond the capacity of their technology. (NHC)*

Pacific. Later, they were to play key roles in pursuing the concept and securing the authorization for the *Albacore*.

Part of *Albacore*'s heritage is her name. The first *Albacore* was laid down at Groton, Connecticut, on April 21, 1941, launched on February 17, 1942, commissioned on June 1 of that year, and lost with all hands on November 7, 1944. *Albacore*'s story spans the Pacific campaign: the early failures, discouragement, deadly danger, the growing successes and mastery, the final crushing blows that finished the Japanese navy as a fighting force.

On her first war patrol in 1942, out of Pearl Harbor to Truk, *Albacore* was hunted and depth-charged for seven hours. She escaped only by absolutely silent running, with off-duty men lying in their bunks and all men standing watch in their bare feet. She expended numerous torpedoes, claimed one hit, and was credited with no damage

This view of the USS Drum *shows the tactical concepts US subs used to crush the Japanese navy and shipping in the later stages of World War II. Using radar they pursued Japanese vessels at high speed on the surface at night, and eluded aircraft by day. Powerful gun armaments permitted them to take on many targets in surface slugging matches, including small warships. (National Archives)*

to enemy shipping. On her second and third war patrols, she got the small light cruiser *Tenryu* and the destroyer *Oshio*. Her skipper, Lt. Cmdr. Richard Lake, was credited with aggressive patrols but severely criticized for his shooting—nine hits out of 46 torpedoes fired—by superior officers who still did not accept the fact that the torpedoes were defective. After a fourth war patrol with no successes, Lake was relieved by Lt. Cmdr. Oscar Hagberg.

On her next three patrols, out of Brisbane, Australia, in 1943, she got a single merchant ship after numerous attacks, again earning severe criticism. She was also nearly sunk by a U.S. Army Air Force bomber that caught her on the surface in bright moonlight. *Albacore* crash-dived but its bombs exploded near her bow as she passed 60 feet, plunging the interior of the boat into total darkness. She went down with her main induction valve open, began to fill, and sank like a stone to 450 feet, dangerously below her maximum operating depth of 300 feet. For two and a half hours, Hagberg and his crew struggled to gain

The Taiho, *sunk by* Albacore *during the Battle of the Philippine Sea, by a remarkable combination of skill and luck. During the "Great Marianas Turkey Shoot" Adm. Raymond Spruance fought a defensive battle against Japanese carrier air, while US subs inflicted most ship losses.* Taiho *is arguably the single most valuable ship sunk by a US Navy sub. (Naval Institute)*

control of the submarine as she oscillated between 30 and 400 feet at various angles.

When *Albacore* returned to Brisbane, Hagberg was relieved by Lt. Cmdr. James W. Blanchard. On her next patrol, she attacked a convoy in company with two other submarines and sank the destroyer *Sazanami*. *Albacore*'s ninth war patrol took her to an area directly west of the Mariana Islands, one of 42 submarines dispersed throughout the central and western Pacific as cover for the landings on Saipan. This thrust took the Japanese navy by surprise, but they quickly saw it as an opportunity to engage the U.S. Navy in a decisive battle. Admiral Ozawa hoped to find and strike the American aircraft carriers first, cripple them, and close the Marianas landing forces with his surface fleet.

But from the moment it left port, the Japanese fleet was shadowed and reported by the U.S. submarines. The American commander, the cautious Raymond Spruance, withdrew eastward to cover his landing forces, fighting a purely defensive battle against Ozawa's massive air strikes. U.S. planes shot down 330 of the 430 Japanese planes sent against the American fleet in the "Great Marianas Turkey Shoot," which effectively destroyed Japanese carrier air for the remainder of the war.

In this Battle of the Philippine Sea, submarines not only provided far more distant reconnaissance than carrier air, but also inflicted the main ship losses on the enemy, foreshadowing the role envisioned for the *Los Angeles*–class nuclear boats in the 1980s. At 0750 June 19, Blanchard raised his periscope and found himself directly in the path of Ozawa's fleet. Suddenly, a large carrier appeared bearing right down

Mrs. Doris Stanton Jowers, the first enlisted man's wife to sponsor a navy ship, about to christen Albacore. *Her husband, Chief Motor Machinist Mate Arthur L. Stanton, went down with the first* Albacore *when she was lost with all hands after striking a mine off Hokkaido, November 7, 1944. Chief Stanton had lost several fingers in an accident in an earlier patrol; he could have taken a medical discharge but chose to stay aboard the* Albacore. *(Courtesy of Mrs. Anneta Stanton Kraus)*

on him. Surrounded by Japanese ships, Blanchard coolly maneuvered to avoid a destroyer and let her approach. As he raised his periscope to take his final bearing and shoot, with the carrier racing by at 27 knots, his TDC—torpedo data computer—broke down.

Blanchard fired all six of his bow tubes without any mathematical solution, using his seaman's eye, then took her deep as the destroyer raced toward him. Pounded by depth charges, he heard only a single definite hit, and Blanchard had no hope that he'd sunk the carrier.

Aboard the *Taiho,* Ozawa's flagship and Japan's newest and best fleet carrier, no one thought the damage was serious either. She was doomed only by an appalling error by her damage control officer. He turned on all ventilators, hoping to disperse the fumes from ruptured gasoline tanks, but only succeeded in spreading them throughout the

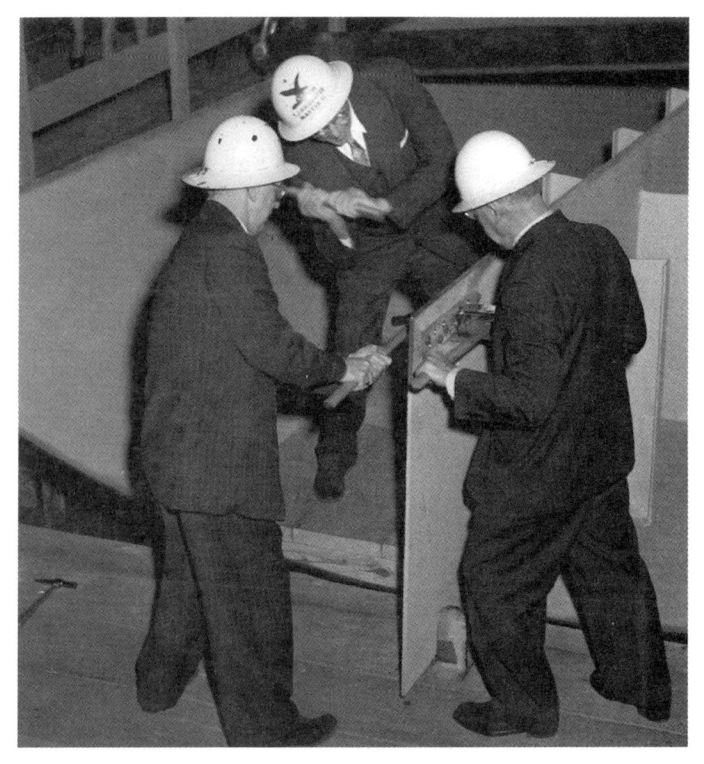

Albacore*'s keel laying, March 15, 1952. Three PNS master mechanics (with a total of 131 years shipbuilding experience between them) drive the first rivet. (Courtesy of Capt. Gummerson)*

ship. Inevitably, a spark ignited this floating time bomb, turning her into an inferno. Months passed before the United States even knew she was lost; Blanchard received the Navy Cross.

Albacore's next two patrols took her to Japanese waters. She left Pearl on October 24, 1944, refueled at Midway, and was never heard from again. According to Japanese records, a submerged submarine struck a mine off Hokkaido on November 7. On December 21, *Albacore* was presumed lost, and her name was stricken from the Navy List on March 30, 1945.

One of the men who went down with her was Chief Motor Machinist Mate Arthur L. Stanton, Chief of the Boat, or senior enlisted man. He had lost a couple of fingers when a hatch slammed shut on his hand during a dive, and could have taken a full medical discharge, but chose to continue serving aboard the *Albacore*. His wife, Doris, just 22 and without much education, was left a widow with small children. A

With "I christen thee Albacore,*" then "I did it! I did it!" Mrs. Jowers sends her down the ways. Her daughter, now Anneta Stanton Kraus, at left, was maid of honor. (Capt. Gummerson)*

person of grit and determination, she went to hairdresser school and supported her two children alone until she remarried.

On August 1, 1953, Doris Stanton Jowers christened the second *Albacore*. Her daughter, Anneta Stanton Kraus, who served as maid of honor at the christening ceremony, believes her mother was the first wife of an enlisted man to sponsor a Navy ship. *Albacore* was the 120th submarine built at Portsmouth. Her keel was laid on March 15, 1952; three master mechanics (with a total of 131 years of shipbuilding experience among them) drove the first rivet. Master Shipfitter and Boilermaker Thomas Gamester drove it, Master Shipwright and Joiner Elmer Pruett was holder-on, and Master Sheetmetal Worker Harold Robbins was inspector. Capt. Edward Craig, commander of the shipyard, kidded that he hoped the rivet would not be rejected.

Seventeen months later, at 5:25 P.M., *Albacore* slid down the ways. Mrs. Kraus remembers the christening ceremony, "a bright beautiful day, with people polite and gracious, all in uniforms." She remembers shipyard commander Capt. Robert Cronin, tall and lean Admiral

Albacore *joins the waters of the Piscataqua, to become a living ship, August 1, 1953. (Mrs. Kraus and Capt. Gummerson)*

Momsen, "a famous person with all of his inventions. . . . everyone deferred to him. . . . he took time to make sure that Mother and I were comfortable." Another of *Albacore*'s main proponents was there, Vice Adm. Edward L. Cochrane, former chief of the Bureau of Ships, then dean of M.I.T.'s School of Engineering.

In his brief address, Admiral Momsen recalled Jim Blanchard and the first *Albacore,* the role U.S. submarines played in the destruction of Japanese seapower, and the tremendous effort involved in defeating Germany's submarines. The new *Albacore* "was born in my office when I held the position of Deputy Chief of Naval Operations four years ago" and would provide Navy ASW officers with "the stiffest competition possible," he told the group. "*Albacore* will open the door to new submarine performance, and by doing so will better prepare us for the underwater menace that might confront us in the years to come."

Mrs. Kraus remembers her mother worrying because she heard that it was bad luck not to break the champagne bottle on the first try, and slinging it with all her might. With "I christen thee *Albacore,*" then "I did it! I did it!" She recalls, "The ship going down the ways, the most gigantic object, the band playing, everyone cheering, it was just wonderful." Then photographs of everyone in front of the *Squalus* memorial. Mrs. Margaret E. Batick, whose husband was lost aboard *Squalus* during her sea trials off the Isles of Shoals in May 1939, presented Mrs. Jowers with the traditional gift of a silver bowl from the shipyard employees.

Some 300,000 man-hours of labor at Portsmouth had made *Albacore* a reality. But another story—that of the conception, gestation, and birth of the *idea* of the *Albacore*—lay five years or more in the past.

Mrs. Jowers photographed at the memorial for the Squalus, *which went down off Portsmouth in 1939, with Vice Admiral Charles Bowers Momsen. Admiral Momsen played the major role in inventing the diving bell used to rescue the* Squalus *survivors. Later, as the first Assistant Chief of Naval Operations for Submarines, he was a leading force in getting the* Albacore *authorized and designed. (Mrs. Kraus)*

II Origin of the Albacore

ON DECEMBER 15, 1953, REAR ADM. CHARLES B. MOMSEN SPOKE at the commissioning of *Albacore.* This speech summarized the basic characteristics and purpose of *Albacore,* and confidently predicted future submarine speeds of better than 50 knots. This last remark was quoted by foreign naval analysts for the next four decades and obviously rocked the Soviet navy, which made vast efforts to achieve such unbelievable results.

Admiral Momsen also said: "Back in 1948, when I held the position of Assistant Chief of Naval Operations [ACNO] for Undersea Warfare, I conceived the idea of designing and building this submarine. . . . In 1948, I held a meeting in BuShips [Bureau of Ships] and asked the design people how they would like to be given a free hand in making a hydrodynamic study of a submarine from the standpoint of submerged performance only. . . . Since it would have no ordnance, only the Bureau of Ships would be involved. We wanted to use conventional power plants, so it could not be called experimental. But since we wanted high-speed, the designers would incorporate in it all of the features of designs which would make a submarine go faster when submerged."

Admiral Momsen was a man of iron courage, a brilliant engineer and inventor, and a superlative leader and administrator. His list of achievements is great: early experiments with submarine-carried aircraft; the invention of the Momsen Lung submarine escape apparatus and the McCann rescue chamber, or diving bell; use of helium in diving to prevent the "bends." He supervised the use of the McCann chamber to rescue survivors of the *Squalus,* which went down on May 23, 1939, during sea trials out of Portsmouth, and supervised the testing off Hawaii in 1943 that discovered the flaws in the design of the American torpedoes. He commanded the first American "wolf pack" and the battleship *South Dakota* during World War II. His appointment as the first ACNO for Undersea Warfare in June 1948 was truly a defining moment for the U.S. submarine service, ending the period of doubt and demoralization following their "silent victory" and the end of the opponent it had been created to defeat, the powerful Japanese surface navy. Under his tenure, the Submarine Service reinvented itself to deal with the Soviet undersea threat. Plainly Admiral Momsen had thought in terms of submarines with *Albacore*-like capabilities for many years ("When in 1939, I suggested that we needed a submarine that could go to 1,000 feet and make 20 knots submerged, it was not taken seriously"), and he played a key role in pursuing and obtaining the authorization for *569*.

But was he the first to conceive of *Albacore*? Or its most important proponent? Capt. Frank Andrews, the submarine project officer at the David Taylor Model Basin (DTMB) from 1953 to 1954, says the concept of a submarine designed for maximum submerged performance—including the "body of revolution" hull, single screw, and the use of HY-80 steel—was first proposed by the Committee on Undersea Warfare of the National Academy of Sciences (NAS) in 1948. The NAS, the equivalent of Britain's Royal Academy, was founded in 1863 to provide the federal government with expert scientific advice from the nation's scientific community. The first problem it was called on to solve was how to compensate for the error caused in the magnetic compasses by the iron hulls of the Navy's new warships. Over the years the NAS created a huge number of committees to deal with specific problems, from insect control in Micronesia to navigation and astronomy for the Navy. The Committee on Underwater Warfare (CUW), however, was initiated directly by the scientists themselves.

World War II antisubmarine warfare, depending on the creation of artificial electronic senses, radar and sonar, to penetrate the submarine's cloak of invisibility, required a huge and innovative research effort, also producing the magnetic anomaly detector (MAD), sonobuoys, the homing torpedo, and operations research analysis. In 1943, when it was apparent that the U-boats were beaten, the effort was switched to support U.S. submarines, providing them with new sensors, weapons, and materiel. But in 1945, with the discovery of German advanced submarine technology, it was apparent that the problems of undersea warfare in the future were far from solved.

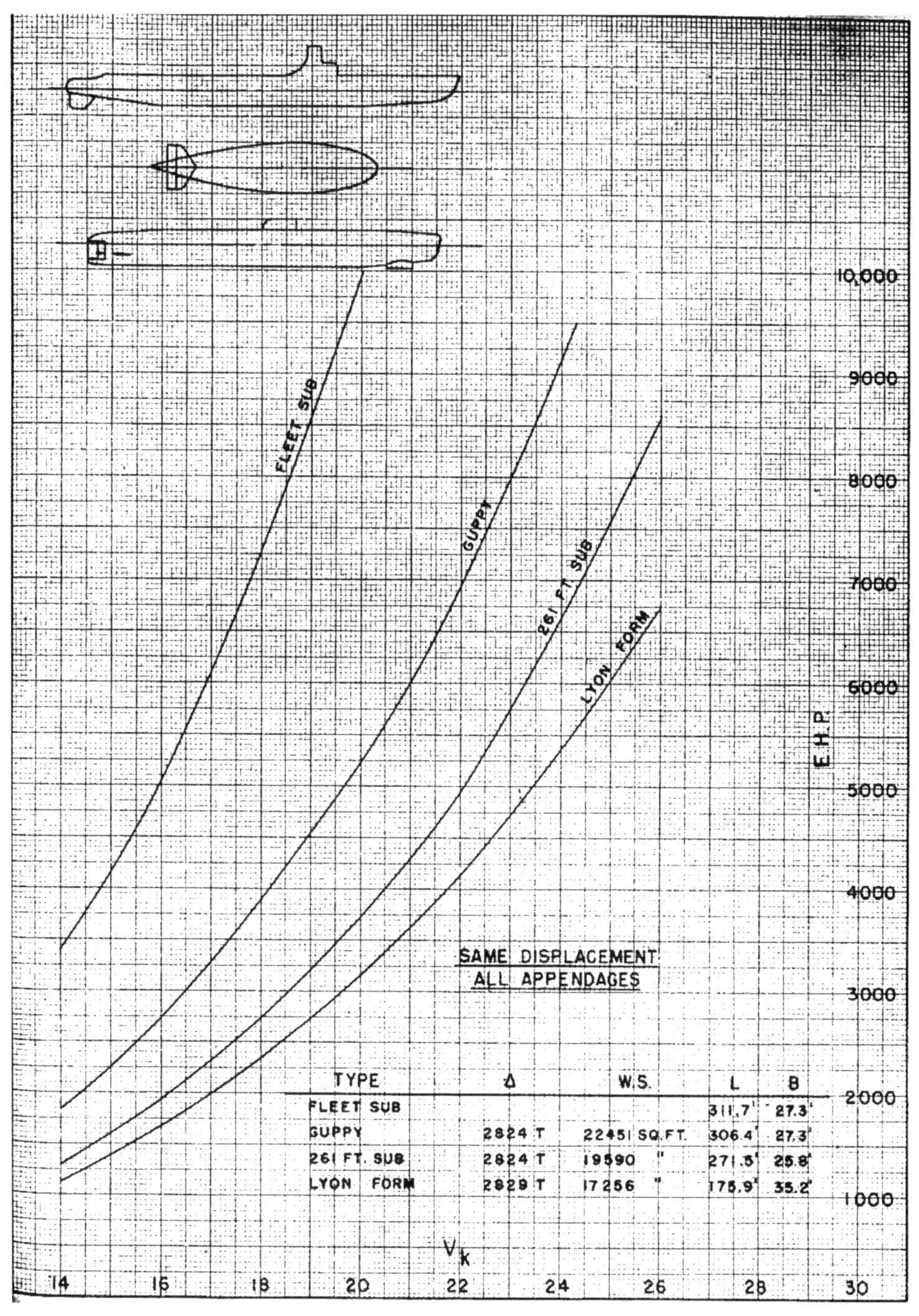

TYPE	Δ	W.S.	L	B
FLEET SUB			311.7'	27.3'
GUPPY	2824 T	22451 SQ.FT.	306.4'	27.3'
261 FT. SUB	2824 T	19590 "	271.5'	25.8'
LYON FORM	2829 T	17256 "	175.9'	35.2'

The "Interim Report of the Committee on Undersea Warfare Panel on the Hydrodynamics of Submerged Bodies" presented the data supporting a wholly new approach to sub design in voluminous detail. This graph shows the horsepower requirements at various speeds of the fleet sub, Guppy, *the 261 foot sub (essentially the* Tang, *an American sub class based on the Type XXI approach of partial streamlining, and the "Lyon form" teardrop hull submarine.*

Leading scientists, including Dr. Gaylord Harnwell of the University of California and Dr. Detlev Bronk of Cornell, chairman of the NAS's National Research Council, sought to continue the close partnership of the scientific community and the U.S. Navy, developed during the war, through a formal liaison body. (As Dr. Harnwell said, "We spent four or five years learning to get along with the Navy—let's not let that disappear.") The Committee on Undersea Warfare (CUW) was established on October 23, 1946. John Tate, of the National Defense Research Council, became first chairman; Harnwell, vice chair, and John S. Coleman, of Penn State, executive secretary. On July 20, 1948, the Chief of Naval Research requested that CUW create a panel to investigate the hydrodynamics of submerged bodies. On November 7, 1949, the committee submitted its "Interim Report of the Committee on Undersea Warfare Panel on the Hydrodynamics of Submerged Bodies." This 64-page report examined the scientific principles governing submarine performance and strongly suggested that the Navy design and build a high-speed research submarine capable of exceeding 20 knots submerged. The report was placed on Admiral Momsen's desk on January 10, 1950, and it played a significant role in securing authorization for *Albacore*. Plainly, though, the idea had already been growing for several years in the CUW, in BuShips, in the submarine community, and at David Taylor, gradually taking on a more concrete and detailed form.

In *U.S. Submarines Since 1945,* author Norman Friedman notes that BuShip's High Speed Submarine Program "began in the spring of 1946. He writes that BuShips officially requested that DTMB undertake the "Series 58" tests, which went on to produce *Albacore*'s hull form on July 8, 1946. (The tests began in July 1949, however.) BuShips had a hand in the development of *Albacore* from the beginning: Some of its key contributors included naval architect John C. Niedermair, "father of the LST"; Vice Adm. Edward L. Cochrane, chief of BuShips and later dean of engineering at M.I.T.; and Rear Adm. Andrew L. McKee, also a member of the CUW Panel on Hydrodynamics. McKee was the designer of the fleet submarine and later served as design director at the Electric Boat Company in Groton, Connecticut, responsible for the design work on most new submarine construction until 1961. Dr. Gary Weir quotes a colleague at Electric Boat, Henry J. Nardone: "[McKee] was one of the last of the breed of engineering duty officers who could sit down and design a submarine almost from scratch."

Dr. Kenneth S.M. Davidson, one of the strongest and earliest proponents of the Albacore, *and founder of the ship research facilities at Stevens Institute of Technology at Hoboken, NJ. Here a younger Dr. Davidson explains about 1935 the apparatus for towing a model hull, here a sailing yacht, perhaps a predecessor of his* America's *Cup-winning* Ranger. *(Stevens Institute of Technology)*

Plainly many people saw *Albacore* as an idea whose time had come and were determined to make it a reality. Just as plainly, there were many who did not, in a Defense Department dominated by the Air Force and concepts of strategic bombing, in a Navy dominated by carrier air, and in a submarine service formed by the experience of World War II, hunting Japanese shipping in wolf packs on the surface at night, in fleet boats with powerful gun armaments and high surface speed. As John Coleman, first executive secretary of the CUW and a leading figure in wartime sonar research, put it: "We encountered much inertia, some hostility . . . the CUW was ignored above a certain level." But, he said, "the *Albacore* was built because enough good men decided it should be built."

Here Dr. Davidson examines a hull model with two longtime Stevens colleagues, Hugh MacDonald and Alan Murray. Stevens not only contributed greatly to Albacore*'s hull, but also that of the* SS United States, *here pictured on the wall. This gigantic liner is officially credited with breaking 44 knots on her trials but reportedly exceeded that by a phenomenal margin, making her undoubtedly the fastest conventional displacement-hulled ship in the world today. (Stevens I.T.)*

Thus, the roles of these key individuals and institutions are overlapping. Perhaps the simplest approach is to look at the contribution of each institution separately, remembering that what is happening is being done by a small group of people working closely together. But—one last time—was anybody *first* to propose the *Albacore*?

When the question was put some 45 years later to John Coleman and George Wood, first and third executive secretaries of the CUW, respectively, both men looked at each other and laughed. George Wood said, "The Navy built *Albacore* to get Ken Davidson off its back."

Gary Weir quotes Davidson's July 26, 1946, letter to Capt. Harry Saunders at DTMB urging the *Albacore* idea. He called for the production of a completely new approach to submarine design, "a rational design" instead of "ceaseless modification and juggling" of existing designs, yielding "a second rate answer." But according to Coleman and

Stevens Towing Tank today. Dr. Davidson was a pioneer in the use of scale model hulls to measure the hydrodynamic resistance of ships, building on the work of such pioneers as Froude in Britain and Admiral David Taylor at the Washington Navy Yard. These men transformed the designing of ships from an art (or mystery) to a science. (Stevens I.T.)

Wood, even earlier, "back in the beginning," Davidson had approached Admiral Cochrane to work out a strategy to get her built: "The Navy wouldn't do basic research on hydrodynamics; David Taylor wasn't interested."

Dr. Kenneth S.M. Davidson was professor of engineering at the Stevens Institute of Technology in Hoboken, New Jersey, as well as director of Stevens' Experimental Towing Tank (now known as Davidson Laboratories). Dr. Davidson was also chairman and a vocal spokesman for the CUW's Hydrodynamics Panel and chairman of committees on towing tanks and hydrodynamics for the Society of Naval Architects and Marine Engineers. In World War II, he worked on the hydrodynamics of seaplanes, PT boats, and torpedoes, and a complex

The term "Lyon form" refers to British physicist Hilda Lyon who studied the wind tunnel tests of airship models to work out the ideal streamlined hull form for the 1929 British airship R-101. This hull was used as the starting point for David Taylor Model Basin's "Series 58" tests which produced Albacore's *hull. (Courtesy of Knapp O'Grady, from the Lord Ventry collection)*

study of the maneuvering capability of many types of ships. It was the study of torpedoes that convinced him that submarines should be similar streamlined bodies. Dr. Dan Sawitsky, Professor Emeritus at Stevens says: "[Davidson] approached the Navy's Experimental Model Basin [predecessor of DTMB at the Washington Navy Yard]—they weren't interested. He spoke to a number of officers—couldn't break through—then one said to go ahead, do model tests. The data confirmed what Ken had been saying. When things started going well, the Navy opened up its facilities and we kept on working. The genesis of *Albacore* is Stevens, with Ken Davidson's leadership and his team."

John C. Niedermair of BuShips, who has been described as primarily responsible for the basic design of almost all naval ships at that time, said: "Ken Davidson suggested the streamlined hull form to me and to others just as though we'd never thought of it . . . one thing he did anyway, he got the top guys to listen to us about it, he did that all right . . . I went up to Electric Boat Co. . . . saw a model of *Plunger* [an 1897 design by John Holland]. I asked if I could have it to show the streamlined sub wasn't anything new."

When the Interim Report appeared, it was voluminous, filled with data, equations, and graphs. But one thing made perfectly clear is that there was no mystery about the potential of streamlining for vastly decreased resistance and increased speed. Drag is essentially based on two things: eddy turbulence caused by the interruptions of flow lines of water around the body, and skin friction. The rounded bow and tapered stern of *Albacore* (as well as the elimination of all projections except the sail and control surfaces) greatly reduced resistance from eddy turbulence.

Early on, Ken Davidson realized that while almost no work had been done on the hydrodynamics of submerged submarines, much had been done in the field of aerodynamics. Air and water differ in density, but the principles of fluid dynamics remain the same for both mediums.

The real problem, however, was "the far-reaching consequences of increases in speed," with their profound and little understood effects on stability, control, and handling. Controlling a submarine moving at the speeds made possible by streamlining and revolutionary new power sources was the problem. And the single concluding recommendation of the report is that the Navy build a test submarine to solve it.

Obviously, proving something on paper does not ensure that it will be done. John Coleman speaks of his own efforts to convince the Navy to build a test submarine, beginning two years before the Interim Report was issued, efforts that were strongly supported by Admiral Cochrane and Adm. Charles W. Styer.

Admiral Styer was Admiral Momsen's predecessor. Carrier air dominated the Navy command; CNOs Nimitz and Denfield had refused to create a Deputy CNO for submarines. Styer and Coleman frequently encountered the argument that the *Guppies* were already 98.6% submarines and any improvement would just add a fraction of the remaining 1.4%.

John Coleman attributes much of the "development strategy" for *Albacore* to Admiral Cochrane. A deeply respected, distinguished naval architect and scientist (and "a hell of a nice guy—very forceful"), Cochrane served the purpose of "verifying the concept. If he said it could be done, then it could," says Coleman. Equally important, Cochrane foresaw that to genuinely convince the Navy of the value of

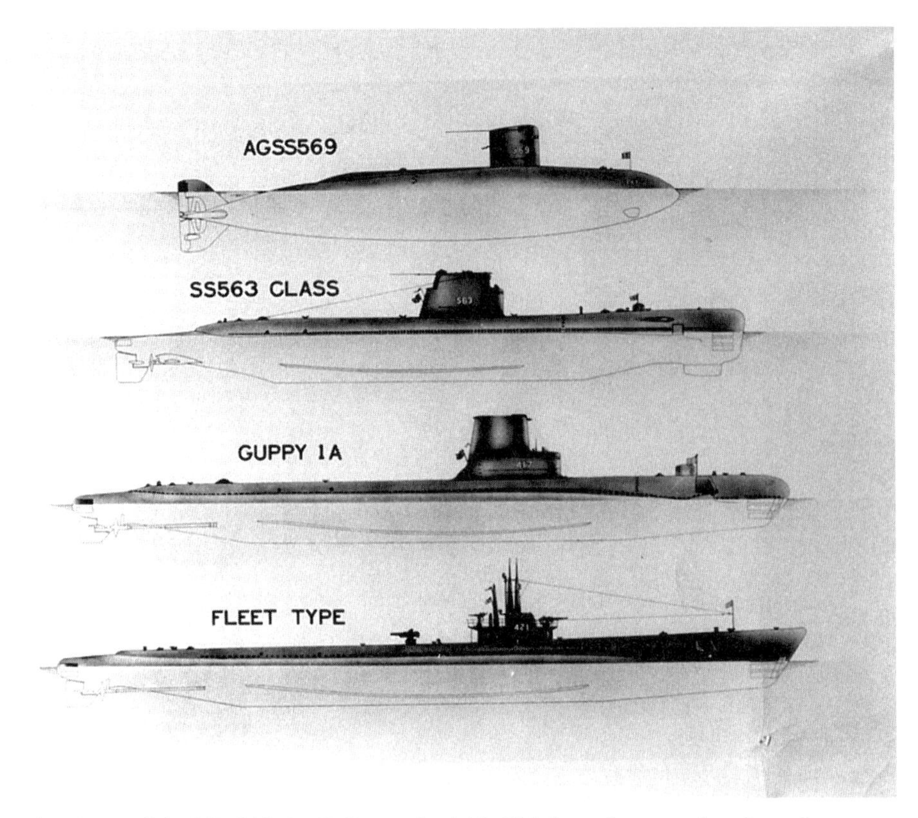

The fine lines of the World War II fleet sub yielded high surface speed and good sea-keeping for the ideal "submersible". Fleet boats were streamlined as much as possible to produce the Guppies *with much better submerged speed. The* Tang *represents the Type XXI approach, a conventional hull streamlined as much as possible and shortened for improved maneuverability. The* Albacore *represents a totally new approach; the pure submarine, designed from scratch. The short, beamy* Albacore *has much the same displacement as the other three craft. (Portsmouth Submarine Memorial).*

Albacore's achievements, it had to be a "ship of the line," a full-size, Navy-manned submarine built by BuShips, not by the Office of Naval Research. Funding for the submarine had to come from the regular ship building appropriations, not from research funds. If *Albacore* were built as a small test vehicle, manned by civilian technicians, she might have been able to prove the theory perfectly, but unless she went head-to-head against the fleet, a submarine dramatically outperforming all other real submarines, would the Navy really appreciate *Albacore*'s revolutionary but highly practical significance? The concept of using *Albacore* as a high-speed target vessel for ASW training was also a useful argument in getting her built as a full-size submarine. Her completion without armament was the other side of the coin; this would ensure that she remain available for research, and not be taken away for operational missions requiring features that could spoil *Albacore*'s hydrodynamic perfection.

As John Coleman says, his own task and that of the CUW was largely one of taking political initiatives within the Navy, convincing it of the vital submarine warfare challenges the service faced—and of the solutions.

The creation of an Assistant to the Chief of Naval Operations (ACNO) for undersea warfare, OP-31, in June 1948 ended this state of affairs. Adm. Forrest Sherman, CNO in 1949, restored ASW and submarine warfare to prominence. The minutes of the Submarine Officers' Conference (SOC), created in 1925 to provide Washington with the views of submariners themselves, now began to show a proliferation of new ideas, designs, and prototypes under discussion. These included the nuclear boat, other air-independent propulsion systems including hydrogen peroxide, the so-called "high-speed submarine," which became the *Tang,* radar pickets, a submarine oiler, a submarine mine layer, a 25-ton midget submarine, a 250-ton boat to test the threat of the numerous Soviet *Malyutka*-class coastal submarines—and *Albacore*.

On January 10, 1950, the Interim Report reached Admiral Momsen's desk, and on January 24, SOC minutes refer to the "SST—experimental hull for studying stability and control at high submerged speeds." On April 3, 1950, the minutes add, ". . . it is to be constructed under a 1950 supplementary ship building program, if . . . approved by Congress."

Meanwhile, at DTMB, several years of work by Dr. Louis Landweber and the Hydrodynamics Division was nearing completion. His team of scientists made seminal advances in the understanding of frictional resistance, surface wave effects, dynamic stability, viscous resistance, computer modeling, and many other areas. In July 1949, Landweber and Morton Gertler began the "Series 58" program, testing 24 lathe-turned, wooden 9-foot models of varying length-to-beam ratios and nose and tail shapes, starting with a form based on the R-101 and H.B. Freeman's study of the Navy's 1931 airship *Akron*. The results of the Series 58 tests showed that a smoothly tapered hull with a length-to-beam ratio of 6:8 and a single screw was ideal.

Later, the National Advisory Committee on Aeronautics (NACA, the forerunner of NASA) and the California Institute of Technology

DTMB scientist Morton Gertler with Dr. Landweber ran and analyzed the "Series 58" model tests in the late 1940s which produced the ideal streamlined hull form for the Albacore. *Here (March 1, 1956), Gertler and Instrument Maker Carson Caudle prepare one of the later "tethered" models of the* Albacore *used to study her behavior while maneuvering at high submerged speeds. (NHC)*

developed free-running models of *Albacore* to test computer predictions of the hull's behavior.

Meanwhile, on March 2, Capt. F.X. Forest reviewed progress at DTMB following a visit by Admiral Momsen. Forest noted: "There has been some tendency to consider the submarine [SST] as an underwater airplane. This analogy is good but . . . limited . . . principally because the airplane is not limited to vertical movements within three of its own lengths." Also, surface effect forces would be 5 or 6 times those of *Guppies,* "Making near surface operation a problem." The new submarine could be driven at 27 knots with less than 10,000 horsepower, and "it is clear that it would be totally impossible to drive the Guppy at 27 knots with any such power."

He noted too: "Perhaps the most pressing and different problem in

A 30′ by 6′ fiberglass model of Albacore's *final design, being tested in the National Advisory Committee on Aeronautics (NACA, later NASA) wind tunnel at Langley AFB. (Naval Institute)*

the entire program is the study of control and response. . . . the submarine in a dive has little or no margin . . . in a dive at 27 knots, the controls must start the pull-out almost as soon as the submarine has entered her dive."

Meanwhile, on March 10, 1950, Secretary of Defense Louis Johnson approved the Secretary of the Navy's request to construct the SST in fiscal year 1951, as a substitution for one of the DDE (escort destroyer) conversions in the 1950 budget. On March 27, the Ship's Characteristics Board (SCB) submitted its "First Preliminary Characteristics for Shipbuilding Project No. 56, proposed for the 1952 increment."

From *Albacore*'s original size of 150 feet long with a beam of 30 feet, a crew of four officers and 36 men, she was enlarged to 200 feet in length, a 27-foot beam, 1,692 tons of surface displacement, with a complement of five officers and 52 men, along with seven scientists. This increase resulted in some loss of speed. Capt. Harry Jackson says that it was originally intended to operate the submarine from Portsmouth, returning to port daily with a very small crew. It was realized that the nearest waters deep enough for submerged testing, the Wilkinson Deep, were far enough from Portsmouth Naval Shipyard to require operations on the basis of weekly cruises and thus enlarged berthing and galley spaces.

On November 29, 1950, the SCB noted that "the Committee on Undersea Warfare of the NAS strongly advocates the construction of this ship." The recommendation was then made that "the Experimental and Target Submarine . . . tentatively designated 'SST'. . be classified as 'AG(SS)' and assigned the name 'T-1.' "

On December 6, 1950, a memorandum from the Assistant CNO for Undersea Warfare agreed that the test submarine be classified as an auxiliary type: "It is felt, however, that the unique features of the ship should be identified in her designation, and that the ship should bear a name. It will be noted that a priority list for naming of new construction submarines was established . . . and that 'Albacore' is the next name on the list.

"It is therefore recommended that: (a) subject vessel be classified AG(SST)-1; (b) subject vessel be named *Albacore*. Signature: C.B. Momsen, R. Adm. USN, ACNO (Undersea Warfare)."

III Design, Construction, and Trials

AUTHORIZED AND NAMED, THE NEXT STEP WAS TO DESIGN AND BUILD *Albacore*. She was designed by Capt. Harry Jackson, whose team included his University of Michigan classmate Eugene Allmendinger, later professor of engineering at the University of New Hampshire. *Albacore* was the beginning of Jackson's long and influential involvement in the Navy's submarine program. He was commissioned ensign in 1941 and went to work on the design of the first wartime destroyer class, the *Fletcher* class. He supervised the docking and repair of the floating dry docks located at Espiritu Santo in the New Hebrides and Guam. After the war he went to the Knolls Atomic Power Laboratory in New York, then to BuShips as the design project engineer for the *Guppy* conversions and the *Tang*-class submarines, as well as *Albacore*, becoming, as he says, Andrew McKee's protégé. Later, he worked as the design project officer on the *Polaris* nuclear missile submarine. He then went to Portsmouth Naval Shipyard as design superintendent, later to Puget Sound and then to Groton as superintendent of shipbuilding. Later he served as technical adviser during the search for the lost nuclear submarine *Scorpion*. After his retirement from the Navy, he taught at M.I.T. and did consulting work for the *Glomar Explorer*.

Captain Jackson says that the initial design work on the SST was being done by Ralph Lacey, who felt frustrated by the Navy's opposition and demands for full military capability. At this point, the CUW, led by McKee, in effect threatened to resign. The fortunate compromise—*Albacore* as a full-scale, Navy-manned submarine, optimized for hydrodynamic qualities but unarmed—was accepted.

Albacore *taking shape on the "building ways." Some problems solved during her construction were the welding of the new HY-80 steel of which her hull is constructed, and the casting of the huge "skegs" or outriggers supporting her tail control surfaces on the "Phase I" stern, permitting them to be mounted aft of her prop. (Portsmouth Submarine Memorial)*

Now the process, from preliminary to detail design to ordering materials to keel-laying, began—but not, without a solution to a number of technological and construction challenges, some of which would plague *Albacore* into her trials and beyond.

One problem was caused by the standard requirement that she be able to remain afloat on her ballast tanks with one compartment flooded. This was easy to achieve in the fleet submarine, with its narrow pressure hull, but not so easy in *569*. Norman Friedman says, "Because the hull had so large a diameter, the only way to hold down compartment volume (and thus satisfy the requirement) was to minimize compartment length." This gave *Albacore* her decidedly odd inside appearance, with wide, short compartments so different from the inside look of a fleet boat which resembled a long, narrow corridor.

Her close, heavy bulkheads put *Albacore*'s structural weight up, made machinery arrangement difficult, and made her control room very cramped—an additional argument for automation and one-man control. The heavy bulkheads could not be included in the nuclear boats, so *569* was the last to meet this requirement.

Albacore was also the first submarine to use the new low-carbon STS steel forerunner of HY-80, which had a yield strength of 80,000 pounds per square inch. Beyond the operational value of deeper diving depths for future submarines, there was the thought of buying her an extra margin of safety should she go out of control in high-speed maneuvers, an ever-present and realistic worry. *Albacore*'s preliminary characteristics called for a 700-foot test depth, with a 1,100-foot collapse depth. Her test depth was finally only 600 feet, less than the *Tang* class; her collapse depth was 1,170 feet. But when the feared event took place and *Albacore* went out of control, she went considerably deeper (possibly 1,600 feet), according to her skippers.

Another issue was saving weight: It was harder to give *Albacore*'s large-diameter pressure hull the same strength as the narrow *Tang* hull. To gain the same strength using regular high-tensile steel would have required double the thickness of the pressure hull, 2 inches. In fact, *Albacore* used HY-80 only for her hull; her frames were made of conventional 50,000 psi yield strength high-tensile steel. *Skipjack* was the first submarine to use the HY-80 steel for her frames as well, and still could not take full advantage of its strength; her hull penetrations, piping, and shaft glands, could not take these pressures. This was first achieved with the *Thresher*-class submarines, which had a 1,300-foot test depth (as described by Friedman and others).

Fabrication of the HY-80 steel required new welding techniques and special "low-hydrogen-type" electrodes. But it was approached with great care. On July 28, 1951, PNS was asked to cut 15 small pieces of HY-80 just received from the Carnegie Company for fillet weld tests to be conducted at the Naval Research Laboratories in Anacostia. No problems were encountered, although later the *Skipjack*-class boats and early British nuclear boats experienced major problems with HY-80, developing numerous hairline cracks in their welds.

Albacore's keel was laid on March 5, 1952, and she was launched on August 1, 1953. Shortly afterward, BuShips stated: ". . . the combination of HY-80 and grade 260 [steel] proved to be eminently weldable. Welder mechanics unanimously agree that this electrode 'handles' very easily . . . fewer repairs of butt and seam welds . . . no occurrences of structural cracking of welds or plate material during the entire construction."

If this success with HY-80 was a happy surprise, a number of other critical engineering issues were not—they caused infinite curses, pain, and gnashing of teeth when they appeared and required the expenditure of much effort for their timely solutions. As her first skipper, Capt. Kenneth C. Gummerson, said, they were under much pressure to complete *Albacore* early—her commissioning was moved up six weeks. *Albacore*'s supporters in Washington, above all Admiral Momsen, wanted to see results. But serious problems arose, causing delays that required heroic efforts from BuShips, Portsmouth Naval Shipyard, and the contractors involved.

In a memo dated February 23, 1951, the commander of PNS reminded BuShips that on November 15, 1950, he had been promised the contract plans for *569* by February 15, 1951. He noted that PNS had been asked to begin construction by July 1951. He was told that the plans were complete, except for the bow and stern frames, which were delayed by the need to order steel for the rest of the vessel. BuShips promised the complete set of plans by the 28th and noted that "the primary objective is to build a structurally sound submarine capable of meeting the requirements of the CNO and the BuShips . . . in as short a time as possible. . . . [T]he most important development . . . is in the field of ship control . . . both automatic and manual." Other requirements were to be subordinated to these; "on-the-shelf" components from *Tang* and the SSKs were to be used where possible, abandoning standardization in favor of speed of design and procurement.

Practically the first issue to be considered was the equipment for DTMB's planned testing program. This was *Albacore*'s first mission, planned in great detail before she was designed. The instruments included dynamometers, strain gauges, velocity rakes, TV cameras, floodlights, hydrophones, and provisions for towing with and without the screw. These instruments were needed for studying drag, stability, acceleration, thrust, wake, boundary layer, cavitation, surface pressure, ship noise, screw design, and the monitoring of the mechanical behavior of *Albacore*'s control surfaces under all conditions. Most of this instrumentation was easily installed and easily removed. PNS needed to know space requirements; procurement was expected to take two years. The requirement for a sonar dome foreshadowed infinite headaches.

But the source of endless agony and delay was an innovative dynamometer shaft, which was to be used to measure engine thrust.

This was a six-foot length of steel to be inserted into the 11 1/2 inch-diameter shaft between the main motor and screw. Like torsion bar springs in automobiles, it was intended to twist; the amount would be calibrated precisely to show the torsional force being exerted on its ends by the torque of the motor and resistance of the screw. It was scheduled for delivery to the Naval Boiler and Turbine Laboratories (NBTL) for calibration on December 15, 1952, but it was forged repeatedly by Bethlehem Steel, turning out defective each time. The material proved difficult to machine as well as to forge. At the Fore River Shipyard in Massachusetts, the yard given this job, further inexplicable delays ensued. The correspondence suggests that A.R. Willner, materials engineer at DTMB, was forced to make frequent use of implied threats of death to both Bethlehem Steel and Fore River, trying to get the dynamometer to NBTL by February 1, 1953, and to *Albacore* by April 1, 1953. When it appeared that the dynamometer would not be finished on time, *Albacore* was ordered to proceed with the rest of her shafting—still with no assurance that the dynamometer shaft would be free of defects. This was merely the beginning; the shaft was to play a part in the biggest shock of *Albacore*'s trials—massive shaft vibration, which led to the failure of the prop shaft seal and flooding by the stern.

Meanwhile, a second source of trouble and delay came from the huge steel castings that formed the stern diving plane and rudder supports, or skegs. The illustrations show *Albacore*'s Phase I stern, a conservative approach in that it places the movable control surfaces behind the screw, where the thrust from the screw's wake multiplies the force acting on the surfaces. All U.S. submarines since *Holland* used this approach; Captain Jackson notes that *Holland* was completed with her planes ahead of her screws and that they did not work. Her trials captain, Captain Cable, found that she could not steer a straight course until the planes were moved behind the screw. DTMB urged that placing the control surfaces forward of *Albacore*'s screw would be hydrodynamically cleaner, but there was some fear of loss of control when backing down. In fact, the Phase I stern would provide perfect, even too much, control and an astounding "dogfighting" maneuverability at the cost of some drag (and it was still terrible backing down).

Structurally, the stern caused big problems. The planes were supported by fins like long arms projecting from the hull, forming a cage around her screw, which were subject to the huge bending and twisting forces in maneuvering. The arms also contained the hydraulic cylin-

This uninformative picture shows Albacore's *hydraulic accumulators, providing the power to work her controls, an area where absolute reliability was required. Loss of control in high-speed maneuvering could have sent* Albacore *below her collapse depth in seconds. Harry Jackson says this system was ultimately based on that of Howard Hughes' gigantic 8-engined flying boat, the "Spruce Goose." (Harvey Horwitz)*

ders that actuated the controls, and these could not be accessed for inspection or repair except when the submarine was in dry dock.

Casting the skegs was also a problem. At an estimated weight of 4 tons each, PNS found them beyond the yard's foundry capacity and on September 3, 1952, asked BuShips to find another shipyard that could manufacture them. PNS got a blistering reply, saying that no such casting capacity existed at any of the Navy's shipyards; such jobs were always farmed out. But PNS forwarded the plans for the castings on October 11, 1952, to the Philadelphia Naval Shipyard for casting, hoping for delivery by January 15, 1953, allowing for a completion date of April 30, 1954. The first casting was defective; the furnace roof material collapsed into the melt. Recasting meant a three-month delay. (Philadelphia also cast *Albacore*'s 8,630-pound service and spare props.)

Another control surface issue concerned the bow planes. They were of an unusual design, having two axes of rotation for test purposes. They were rigged in by a double motion, slanting inward and rotating. All motions were controlled by a single set of large hydraulic cylinders that took up much space in the forward compartment. These were noisy, bulky, and largely unnecessary since *Albacore* operated mostly below periscope depth, where these planes would provide precise depth control. They were eventually removed. But these planes needed to traverse much more quickly than the stern planes and fear of a plane casualty was always present.

If the supreme consideration in *Albacore*'s design was submerged performance, the second was the fear of an out-of-control dive, sending her below collapse depth, which indeed happened repeatedly throughout her career. Structural strength, "brakes" to arrest a dive, and ways to prevent control surface failures were important in her design.

Captain Jackson notes that *Albacore*'s hydraulic power system, as well as much of the other technology, was derived from Howard Hughes's giant wooden 8-engined flying boat, the H-4 *Hercules* or "Spruce Goose." Most big bombers used in World War II still employed manual cables to work their rudders and elevators, worked by the pilot's muscles exerted on the joystick. They reached the limit, however, where power control was necessary. The B-29 used electric servomotors, requiring a heavy, bulky onboard motor generator. The postwar B-50, externally similar to the B-29, replaced this control system with a much lighter and more reliable hydraulic system. The 320-foot-long wingspan *Spruce Goose* pioneered the use of hydraulic control. When the original system proved inadequate, Hughes produced a new design with two redundant systems.

Albacore needed high levels of power and safety for her control surfaces, and Captain Jackson sought to give her aircraftlike standards of maintenance and reliability. Her plant consisted of three hydraulic pumps in parallel with two pressure mains running through the ship, the "vital system" and the "main system." There were eight accumulators or pressure reservoirs, each of which could be isolated from the system. It was absolutely essential for *Albacore* to have sufficient hydraulic power in reserve to cycle her dive planes instantaneously from full down to full up, in case of a total loss of power to her hydraulic pumps in the midst of a high-speed dive. This 3,000-pound psi high-pressure hydraulic system was pioneered in the *Tang,* where it gave endless problems; but by the time *Albacore* came around, this system worked fine.

Another constant concern was hull smoothness. When Admiral Momsen had predicted 50 knots submerged speeds—as he continued to do throughout his life—he was counting on a breakthrough in the area of boundary layer control. Referring to Morton Gertler's figures showing 35,000 hp required to drive a 1,000-ton *Albacore* hull at 60 knots, he noted that at that speed skin friction constituted 60 percent of the total drag. Much concern for surface friction was shown throughout *Albacore*'s construction, with the thickness of reinforcing beads over welds versus smoothness being debated at length. Her hull roundness was carefully checked with a profilometer, and a variety of hull coatings—zinc chromate, hot and cold vinyl plastic, for example—were to be tested for their effect on drag as well as maintenance. The CUW suggested filling in dished areas of the hull with auto body filler—"Bondo"—as was the practice with sailing yachts. (Could this idea be from Ken Davidson?)

Smoothness remained a concern. Second *Albacore* CO Jon Boyes turned out his entire crew at one point to sandpaper and polish the submarine's hull in dry dock and reported a gain of 2 knots in speed! Her never-completed Phase V trials involved the injection of viscous polymer fluids into the boundary layer. An example of how seriously this was taken during *Albacore*'s construction was a debate on the use of Scotchlite, a thin, rubberlike material for her name, draft marks, and so on, to eliminate raised plate numerals. BuShips suggested painted numerals delineated with punch marks—but that could have weakened the HY-80 steel.

The snorkel continued to be debated, especially since this affected battery ventilation. (The battery, an Exide Guppy II, was a problem because of the odd shape of *Albacore*'s battery compartments; PNS made a wooden mock-up.) On April 21, 1951, she was directed to be delivered without the snorkel or closed cell battery ventilation, but with "provision made for their ready installation." PNS objected, stating that this constituted a change in the specifications of the boat. BuShips later agreed; this would have changed the size of her sail.

As the construction of *Albacore* progressed, a thousand items were raised and dealt with: The provision of a 1,200-pound stockless anchor; an air horn; flags, pennants, semaphores, union jacks; life rafts; 10 additional bunks for "passengers" (the scientists); control instrument panel from the Arthur D. Little Company; automatic depth control from the Askania Regulator Company; hydraulic operation of the clamshell bridge covers. The deletion of controls from *Albacore*'s enclosed bridge position was also considered (its windows can be seen in early photographs). Calculation of lead ballast and reserve buoyancy was done, and gauges for her deep structural test dives were provided. Could she mount her stern anchor light on her rudder post? Mounting the light on her hull would leave one-third of her length dangerously unmarked.

A full-scale model of a section of her hull was constructed and studied with PNS's brand-new pressure test tank, as well as a series of one-quarter and one-half-scale models.

As completion of *Albacore* approached, Ken Gummerson asked for a Torpedo Bearing Repeater to be mounted, so that compass bearings could be taken off *Albacore*'s periscope. The reply from BuShips blistered his paint: This was a piece of fire-control equipment—*Albacore* was to be totally unarmed! They later admitted that it was necessary; *Albacore*'s cramped flying bridge required the navigator to be down in the control room, getting his bearings via the periscope. Gummerson also requested a scanning sonar, planned for later installation for the target role, to avoid collisions with ships while surfacing. This possibility was a serious hazard; Gummerson was given an old JP sonar which proved to be totally ineffective and was later replaced with a similarly ancient JT sonar (from an old fleet submarine) that earned mixed reviews and much modification.

The Pancake Diesels

When *569* underwent trials, two very serious flaws appeared: the "Syntron Seal," was one, and her engines were the other. These were the infamous GM-16-338 "pancake diesels," so-called because of their 16 cylinders in four stacked radial banks of four. This lightweight, high-speed engine was developed for landing craft in World War II. (Successful? Harry Jackson says: "They *ran*. They had a short life but were said to be satisfactory.") The Navy wanted small engines—high power, low volume—for the *Tang*s to keep their hulls short and maneuverable. Later they were lengthened 30 feet to take in-line engines.

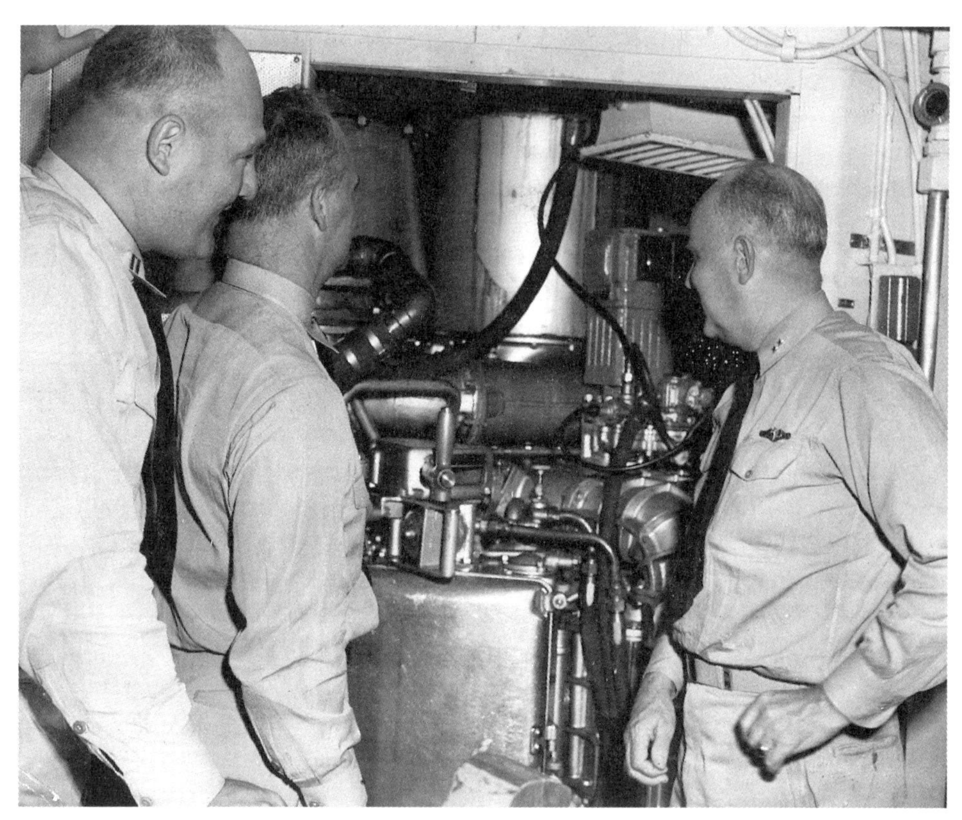

An unprepossessing portrait of the villain in the story, one of Albacore's *GM 16-338 sixteen cylinder radial or "pancake" diesels being inspected by Arleigh Burke and Lord Louis Mountbatten. Of light weight and advanced design, they provided endless breakdowns and headaches, but were the only diesels that would fit in* 569's *little, odd-shaped engine room. Their flaws bedeviled her career and when the supply of replacements and spare parts ran out, her life was over. (PSM)*

They used four pancakes compared to *569*'s two; and because of her size and odd-shaped engine room, they were the only engines she could take. (One compromise to her submerged form they imposed was the need for a short, flat-decked superstructure aft of her sail to house mufflers and exhausts.)

Albacore's surface performance was an eternal source of aggravation. The teardrop hull wasted much power by making a huge bow wave on the surface, and she tended to push her downward curved nose under. Her top surface speed was 14.5 knots, but according to engineering officer Ron Hines, she could do no more than 139 revs, 10 to 12 knots, before her bow wave was up to her sail. With nothing to damp rolling—like bilge keels—her crew suffered miserably even in moderate seas.

Lack of diesel power caused problems too. Her 7,500 hp main propulsion motor could discharge her battery a lot faster than her two 1,000 hp diesels could charge it back up. A full charge was good for 8 to 9 hours of submerged runs, but recharging it took 18 to 20 hours during which she could make maybe 5 knots. Surfacing with an exhausted battery in a hurricane, she began rolling so badly that she started taking on water through her main air induction at the top of her sail. Unable to submerge or make headway, she came close to being lost.

And this is when the diesels were working! Even the *Tang*s, with four engines, often needed to be towed in. Under Capt. Frank Andrews, *Tang* had to be towed back from England. Ned Beach tells of his new command, the *Trigger*, broken down and drifting on her shakedown cruise to Rio de Janeiro. *Albacore* CO Bill St. Lawrence told of losing both engines off Long Island; lacking any suitable padeye, the tow line had to be passed around the sail. But the 16 spare diesels from the *Tang*s provided *Albacore* with a reserve that lasted her entire career. When they were gone, her life was over.

Many COs had choice words for them: "100% headache," or "not worth the powder to blow them to hell with." Harry Jackson explains their weakness as due to light construction: "seven or eight tons more and they'd have been satisfactory and thus maybe standard sub engines today. They would have had better bearings, vibration was fierce, plus the COs drove them too hard, wore 'em out."

Ken Gummerson blamed their small bearing clearances: "Dirt that would go through the in-lines would chew them right up." This and the fact that only part of their lube oil was filtered made a formula for trouble. During *569*'s post-shakedown availability they installed a big filter to purify all the oil. "That was the big correction to the engine design that made it able to go," said Gummerson.

Another flaw was that the generator was directly below the diesel. Design defects permitted leaking lube oil to accumulate around the main shaft seal, which then dripped down and fouled the generator—this was largely solved in the *Tang*s.

But the problems of structural weakness and vibration remained. *569* lost an engine in June 1954; a main bearing cap bolt broke, wiping out the crankshaft and four connecting rod bearings. The fourth casualty to a 338 in three months, it left the supply of spare parts nearly depleted.

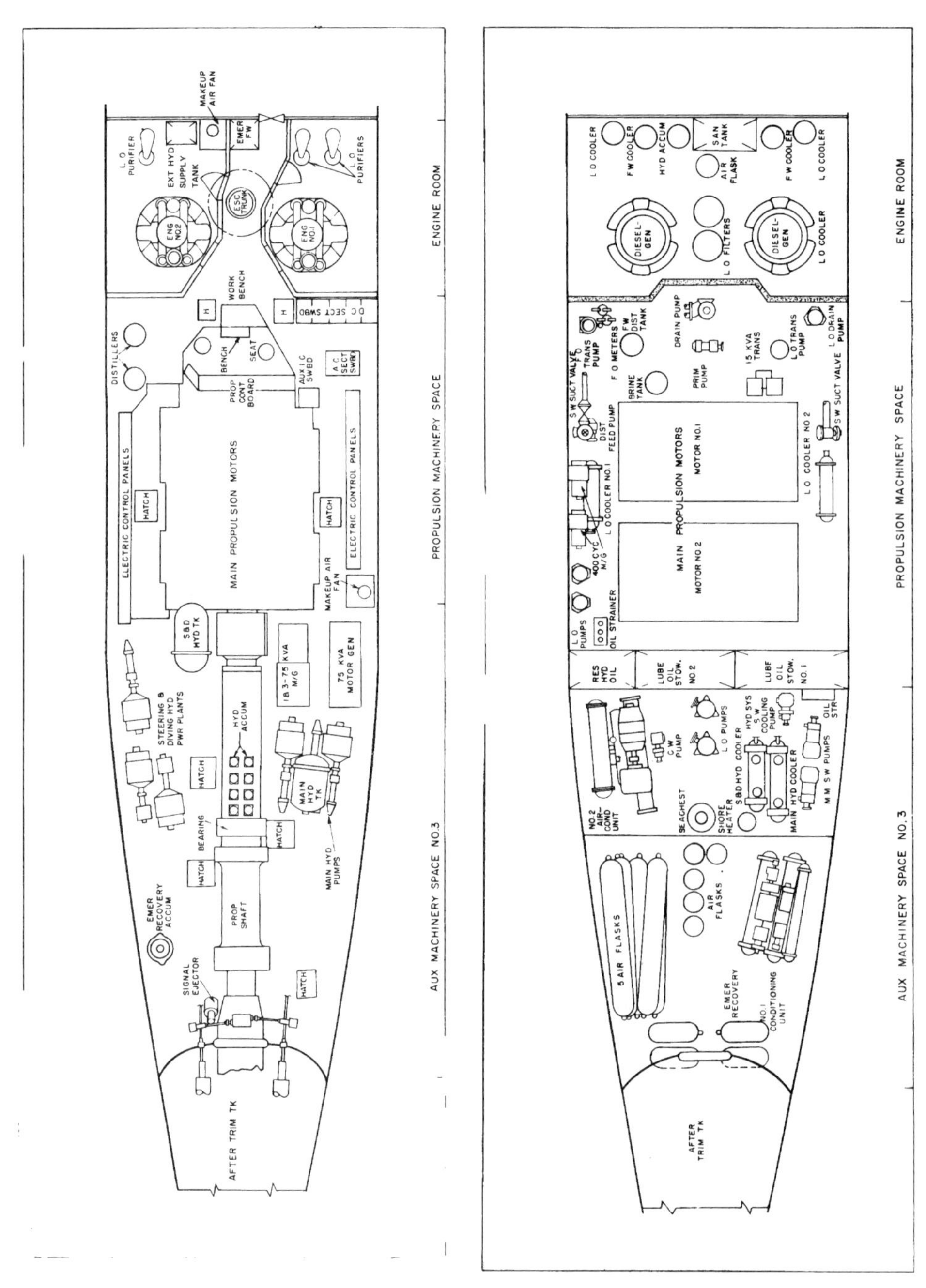

These plans, taken from the "Training Aid Booklet" supplied to all crew members, actually shows Albacore *in her final configuration with two electric main motors, X-stern, and contra-rotating twin props. (Courtesy of Russell Van Billiard)*

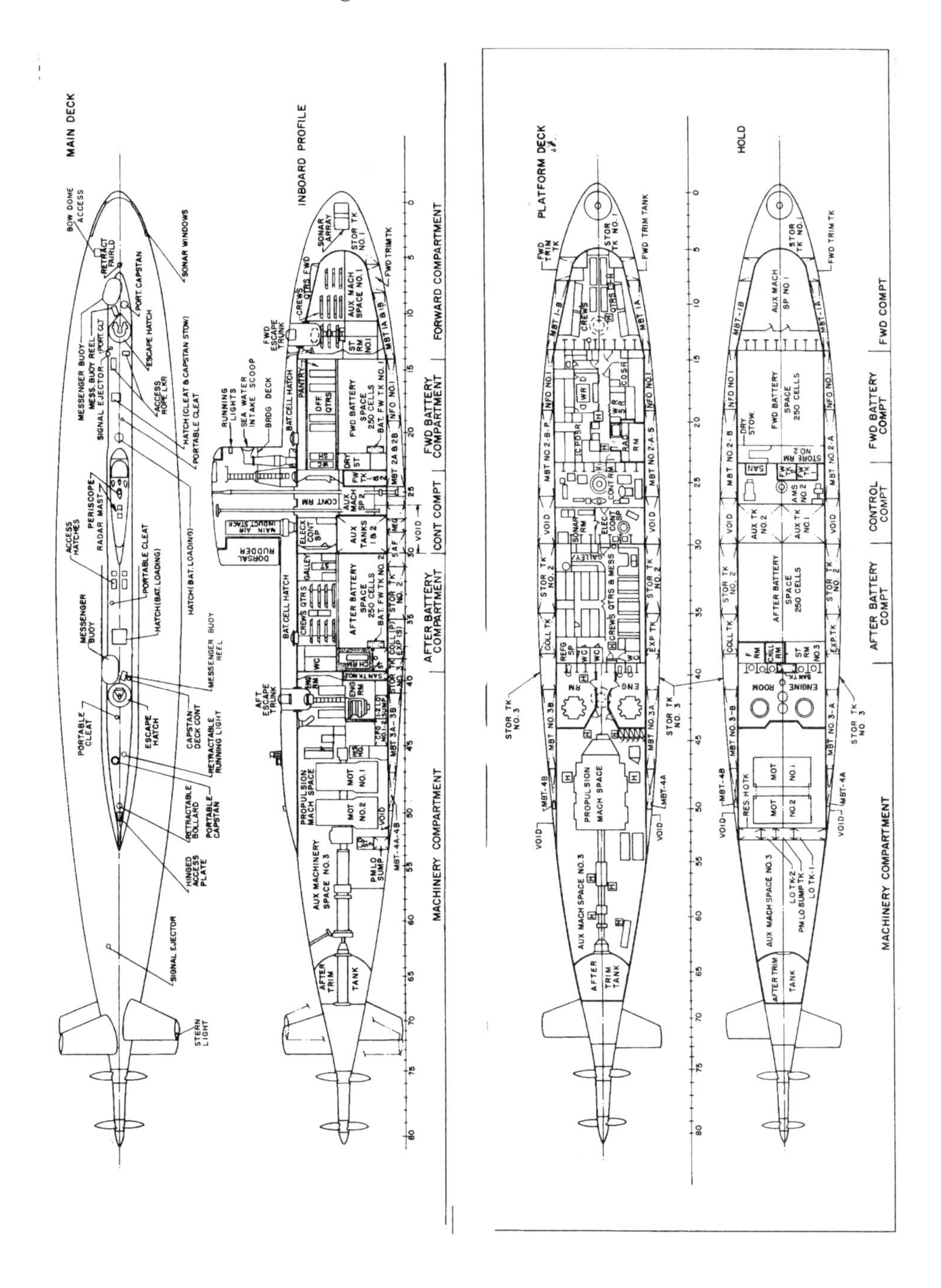
MAIN DECK
BOW DOME ACCESS
SONAR WINDOWS
PORT. CAPSTAN
ESCAPE HATCH
MESSENGER BUOY
SIGNAL EJECTOR
PERISCOPE
RADAR MAST
ACCESS HATCHES
PORTABLE CLEAT
HATCH (BAT. LOADING)
MESSENGER BUOY REEL
CAPSTAN DECK CONT
RETRACTABLE RUNNING LIGHT
RETRACTABLE BOLLARD
PORTABLE CAPSTAN
HINGED ACCESS PLATE
STERN LIGHT
INBOARD PROFILE
RUNNING LIGHTS
SEA WATER INTAKE SCOOP
BRDG DECK
BAT. CELL HATCH
FWD ESCAPE TRUNK
AFT ESCAPE TRUNK
DORSAL RUDDER
MAIN AIR INDUCT STACK
CONT RM
SONAR ARRAY
STOR TK NO. 1
FWD TRIM TK
AUX MACH SPACE NO. 1
OFF QTRS
PANTRY
FWD BATTERY SPACE 250 CELLS
GALLEY
CREWS QTRS
AFTER BATTERY SPACE 250 CELLS
ENG RM
PROPULSION MACH SPACE
MOT NO. 1
MOT NO. 2
AUX MACHINERY SPACE NO. 3
PMLO SUMP
AFTER TRIM TANK
FORWARD COMPARTMENT
FWD BATTERY COMPARTMENT
CONT COMPT
AFTER BATTERY COMPARTMENT
MACHINERY COMPARTMENT
PLATFORM DECK
FWD TRIM TK
CREWS QTRS
SONAR RM
GALLEY
CREWS QTRS & MESS
ENG RM
PROPULSION MACH SPACE
AUX MACH SPACE NO. 3
AFTER TRIM TANK
STOR TK NO. 3
HOLD
AUX MACH SP NO. 1
FWD BATTERY SPACE 250 CELLS
AFTER BATTERY SPACE 250 CELLS
ENGINE ROOM
MOT NO. 1
MOT NO. 2
AUX MACH SPACE NO. 3
AFTER TRIM TANK
FWD COMPT
FWD BATTERY COMPT
CONTROL COMPT
AFTER BATTERY COMPT
MACHINERY COMPARTMENT

Bad as these engines were, though, COs often praised their expert engine room crews and their ability to keep them running. Chief Engineman Stan Zajechowski says their small size eased repairs, with only two or three men needed to pull a piston. He also suggests that some engine room crews were simply ignorant of the maintenance required; this was their first experience of a high-speed diesel running at 1,500 rpm (they were accustomed to the 750 rpm of the *Guppy* diesels). Vibration caused problems unless nuts were tightened to torque and lock-wired. Maintained "by the book," he found that they performed well.

Her main electric propulsion motor seemed cast iron by comparison. But its lube oil pumps were run by AC current off two auxiliary 750 KVA "Little Bertha" motor generators—which had scores of problems. Had they lost both at the same time, that would have shut down the main motor, but that never happened. One 750 KVA was the prototype for the *Tang* sets, used at PNS for tests. Differences between the two sets made their operation in parallel impossible to regulate. And it was necessary to cannibalize the practice switchboard at the Sub School in New London for the instruments and rheostats needed for their control panel; otherwise, it would have meant a six-month delay.

Commissioning

Captain Gummerson's first hurdle (besides getting *569* completed) was INSURV, the Board of Inspection and Survey, which came aboard November 20, 1953, at Berth 11B, in Portsmouth. They found her still not habitable, but 84 percent complete, with dock trials under way. On November 27, 1953, PNS commander Capt. R.E. Cronin informed BuShips and CNO, "AGSS 569 is in a proper condition to be commissioned."

On Sunday, December 5, 1953, *Albacore* was commissioned, with Ken Gummerson, who had seen destroyer action at Okinawa and gone to Sub School postwar, as her first CO. Captain Cronin presented him and his officers with a silver bowl and a pair of candlesticks on behalf of Mrs. Jowers, in memory of Chief Stanton. Admiral Momsen, now Commandant of the 1st Naval District, spoke again, discussing in detail hydrodynamic factors in *569*'s design. The bitter, confusing Korean War had ended recently in an uncertain truce. The Soviet Union had 345 operational subs, including 47 streamlined *Tang* equivalents, mostly the first of 236 *Whiskey*-class Type XXI clones, but including some

December 5, 1953. Albacore*'s commissioning ceremony. She lies at the pier, essentially complete, ready for trials and then a multitudinous variety of scientific tests designed by DTMB scientists before the* Albacore *herself. (Mrs. Kraus)*

Admiral Momsen spoke again at her commissioning. His predictions of phenomenal submerged performance by future submarines rocked the naval world. (Capt. Gummerson)

oceangoing *Zulu*s, big, fast, long-ranged, easily capable of operating off the U.S. coast. The Soviet submarine fleet reached a peak of 473 boats in 1957. The only hope the United States had of countering this force was a revolution in technology—by quality, not quantity. This would come, in the marriage of *Albacore* and *Nautilus*. But it worked; the Soviets planned 160 *Foxtrot*-class improved *Zulu*s but canceled 100 to concentrate on catching up to the American lead in nuclear sub development. As Momsen said: "As a submariner it is a signal honor and a privilege to stand here and see this new *Albacore* breathe the first breath of life . . . a great deal depends on you officers and men who are commissioning this ship today . . . The *Albacore* may lead the way to mastery of the seas by the submersible. In any event, it is important that we retain the initiative in all phases of submarine development; for we may be certain that our enemy will not neglect this field which offers unequaled opportunity to challenge the mastery of the seas."

And here she is! Phase I Albacore *alive, at sea, underway. She shows many early features, Phase I stern, windows for her enclosed bridge, bow planes, no bow sonar dome (and no workable sonar at all, a serious lack.). (Naval Institute)*

Trials: The Testing of the Snark

Now to make her, still unfinished, a living ship. Builder's trials of all equipment, at dockside, were under way. Preliminary acceptance trials by INSURV. Deep-dive structural reserve tests, conducted by DTMB and BuShips. Conditional acceptance, shakedown cruise, post-shakedown overhaul, remediation of PAT deficiencies. Standardization trials on the measured mile off Provincetown. Final acceptance trials with INSURV. All this consumed Ken Gummerson's 1954.

At *569*'s 39th-anniversary reunion, he recalled that after riding the *Albacore,* Ken Davidson sent him a copy of Lewis Carroll's "Hunting of the Snark," telling of the chase of an indefinable creature, on a ship that occasionally sailed backward, "where the bowsprit got mixed with the rudder sometimes," and with a blank chart. Gummerson quotes Lewis Carroll:

> He had bought a large map representing the sea,
> Without the least vestige of land:
> And the crew was much pleased when they found it to be
> A map they could all understand.

This was charming no doubt: but they shortly found out
That the Captain they trusted so well
Had only one notion for crossing the ocean,
And that was to tingle his bell.

"The Bellman—the captain in the story," said Gummerson. "went to sea with some funny things on board, he went to sea with some funny, funny people, and went to hunt a funny, funny thing. The *Albacore* was like hunting the snark. She was a very elusive object. Once you felt you'd mastered its characteristics, poof! It was gone! And you had to learn it all over again."

Gummerson quotes Lewis Carroll again:

Come listen my men while I tell you again
The five unmistakeable marks
By which you may know, wheresoever you go,
The warranted genuine Snarks.

Let us take them in order. The first is the Taste,
Which is meagre and hollow, but crisp:
Like a coat that is rather too tight in the waist,
With a flavor of Will-o'-the-wisp.

"When we came to the *Albacore,*" Gummerson went on, "it was unbeknownst to anybody just what we had our hands on. None of us had ever gone faster than 8 or 9 knots submerged or tilted 30 degrees or 40 degrees up or down—especially down! The first time we submerged Harry Jackson said, We don't know whether it's going to go down stern first or bow first. And I said, Why didn't you tell me this before? And he said, We didn't want you to miss out on any sleep."

When Gummerson first took her out he was told not to exceed two-thirds power. In the fast Piscataqua, with its powerful tidal currents, departure depended on the tide, had to be at slack water, and required a tug and pilot. Going out after midnight, they spotted a red running light; obviously they were heading straight for a ship. Gummerson ordered two-thirds power—nothing happened. Then "Back emergency," and that stopped her just in time. Her first submergence took 30 minutes—versus a fleet sub's 30-second crash-dive.

Her next major evolution was the deep dive—they had to go to the channel off Cape Ann to find the necessary 900 feet. With 48 people on board including the DTMB strain gauge party, nobody had any place to sleep for three days.

Albacore, *April 5, 1954, at sea to have her aerial recognition photos taken, including these unusual angles. (National Archives)*

They had three or four yard workmen permanently assigned to them. "Everything we did," said Gummerson, "they went along with us except the big shakedown. So when we came back into port they knew how it'd react. These were the guys who designed the hydraulic system, worked on the piping, the electricians—they were leading men in the shops."

And then INSURV on board for Preliminary Acceptance Trials, May 3–4, the "final graduation exercise"—which the *Albacore* almost flunked! First, if her bow planes were put on full rise over 18 knots, they stuck. Second, water poured in around her after hatch: "They were wearing rain gear back there," Gummerson said. Third, the Syntron seal.

When they tried backing down at full power, her prop shaft began a tremendous vibration, and water started pouring in around the stern shaft-bearing seal. (DTMB had foreseen this possibility, and designed a set of diagonal struts between the skegs to stiffen her stern. But they would have imposed some cost in speed and drastically increased her turning radius, so it was hoped they could be dispensed with.) The infamous DTMB dynamometer torsion shaft section was partly to blame due to its flexibility. Because of its lateness, a "stub shaft" had been fabricated to take its place. Afterward pulling the torsion shaft—on which so much ingenuity and pain had been lavished—and replacing it with the stub shaft reduced the vibration, lifting it out of the critical rpm range.

Second, it was determined that deflection of the thrust-bearing foundation had occurred. This was solved by reinforcing it—welding a piece of I-beam on each side between the foundation and a frame. Finally, the seal was replaced with a new one as a safety factor; it used 12 internal coil springs to exert pressure on the shaft, 15-pound springs in lieu of 6 pounds.

INSURV authorized acceptance of delivery provided 19 deficiencies were corrected. The Commandant of the 1st Naval District accepted custody of *569* for reduced service as of March 15, 1954. She docked for the modification of the Syntron seal May 17.

Her shakedown cruise to Key West and Havana in June began with a feeling of relief. "This was fun," recalls Gummerson. "We were all by ourselves, no passengers!" The first jolt was her motion in the seas off Cape Hatteras; without bilge keels it was "like a big rocking chair." They had a hard time raising home base by radio, to call in surfacing and diving messages; "Otherwise," he said, "They'd start the wheels for a rescue operation. We surfaced at the end of a day's run off Jacksonville and I couldn't raise the boss. I phoned the marine operator to put through a person-to-person call and the people in New London wouldn't accept the charges!"

Then came the engine breakdown. Gummerson describes it: the diesel "went berserk—we spent three weeks in Key West getting the engine back together. We started back and the other engine went. We were in dire straits keeping that big battery charged. I decided to accept a tow back to Portsmouth.

"We came back for seven months post-shakedown availability. PNS fixed up all the things wrong. We felt pretty good about it, felt like we could operate the boat safely, felt like we were getting familiar with it. We stopped having that 'Snark feeling' all the time."

Then came the stabilization trials on the measured mile off Provincetown. Admiral Lockwood—famous commander of the World War II Pacific sub force—was to be there. But where to find a boat to bring him aboard? *569* possessed no such thing. Another "Snarkish" event. She couldn't anchor. A lobster pot had fouled her chain, which stuck in her hawse hole. She was short an officer (banged up in a skiing accident). Out to sea all night, her captain had to stand watch, so "no more skiing!" The "trials were very exciting," Captain Gummerson said. "They scared me sometimes, and I was fearless."

During this time America had become aware of *Albacore* when Cornelius Ryan (author of *The Longest Day)* spent two weeks in Portsmouth with her doing an article for the April 1955 issue of *Collier's* ("I Rode the World's Fastest Sub" appeared with "Who Can Save Asia?" and "Backstage with Gleason"). A little breathless but not bad at all:

"Gummerson increases the speed again, and again. The motors turn faster in a steadily increasing tempo. On the dial of the speed indicator, the needle eats up the knots . . . hits 20 . . . passes 20 . . .

"'Left full rudder,' snaps Gummerson. As he speaks, he steps down from the bridge and wedges himself into a corner. Caught away from a leather strap, I brace myself between the navigator's desk and the electrical switchboard. We bank almost before Gummerson finishes speaking. There's a sharp jar as *Albacore* heels over. Everybody rolls and sways with the sudden violence of the turn. I bend almost double over the navigation desk. There's a clatter as loose objects almost begin to

Albacore*'s uniqueness and achievements brought her considerable publicity; she was featured in several magazine articles and at least one TV show, Dave Garroway's "Wide Wide World". Perhaps the best treatment was "I Rode the World's Fastest Sub" by Cornelius Ryan (author of* The Longest Day*) in the April 1, 1955* Collier's *magazine.*

slide in their tracks. On the navigation desk a pencil rolls; before it stops we level out again. The pencil hesitates, wobbles, and comes to rest.

"'Right full rudder!' orders Gummerson."

Ken Gummerson came to know and like Ryan personally and laughed off the melodrama. He says *Albacore* gave Ryan a moment of genuine terror. They were at 300 feet, going along very nicely, when there was a tremendous racket. "I think we lost the rescue buoy" (which provides a downed sub with a link to the surface). But what to do now? They locked the screw before surfacing to prevent wrapping the buoy cable around it. It was cold, windy, and 2 A.M.

"I said, 'Somebody has to go back there with a pair of wirecutters and cut that thing loose.'

"The Chief of the Boat—Charlie Simpson—says 'Yeah, I'll do it.'

" 'Don't get lost.' "

Her next skipper, Jon Boyes, did not fare so well with *Police Gazette* writer N.K. Perlow: "My Terrifying Dive in the World's Fastest Submarine." It suspiciously resembles Ryan's article, down to the gee-whiz emphasis on the "pilot's" safety belt and the subway straphanger leather loops. And he also has seamen desperately trying to fix a leak. Apologies to Ted Davis, who still winces over this:

"Lieutenant Ted Davis, who was piloting the vessel, quickly turned to Jon L. Boyes, the Skipper.

" 'Captain, I've lost control! The planes are locked!'

"The muscles in Boyes' sun-soaked face stiffened and he snapped: 'All stop! Cut the engine!'

" . . . the tension thickened. Auxiliaryman Alfred Woodard reported: 'Captain, the hydraulic valve in the plane system is jammed!'

"I stood frozen; we were nose-diving into the floor of the ocean! All the horrors that trapped men faced in sunken submarines flashed through my mind . . ."

If he could have been sent there without *Albacore,* some would have approved.

Albacore underwent Final Acceptance Trials on February 1–2, 1955. On April 12, all work items were complete and deficiencies remedied, and ComSubLant recommended ". . . *Albacore* be accepted for unrestricted service."

Unrestricted indeed; they hadn't seen nothing yet! But as Ken Gummerson, who died in 1994, put it: "This elusiveness of the *Albacore*—throughout her career she always kept that Snarkish quality."

"He remarked to me then," said that mildest of men,
" 'If your Snark be a Snark, that is right:
Fetch it home by all means—you may serve it with greens,
And it's handy for striking a light.

"But oh, beamish nephew, beware of the day,
If your Snark be a Boojum! For then
You will softly and suddenly vanish away,
And never be met with again!' "

IV *Flying the* Albacore

ON JUNE 30, 1955, LT. CMDR. JON BOYES TOOK OVER THE COMMAND of the *Albacore* from Lt. Cmdr. Kenneth Gummerson at Portsmouth Naval Shipyard. With Jon Boyes began perhaps the most crucial and revolutionary period of *Albacore*'s career, her Phase I trials. Though they lasted less than six months, they showed for the first time the potential for naval warfare of a high-speed submarine able to maneuver freely submerged, in three dimensions like an airplane. *Albacore*'s Phase I performance engendered an entirely new conception of the submarine that was debated over the next four decades, repeatedly put forward by her partisans but never fully adopted by the Navy. Indeed, it seems to have left a stronger mark on Soviet submarine design and tactics than on our own. And in some respects it was never equaled: When *Albacore* went into dry dock for her Phase II conversion in December 1958, she lost certain capabilities that no other submarine has equaled.

As *Albacore*'s executive officer from November 1954, to September 1955, Capt. Ted Davis, put it, "Believe me, Jon Boyes made the *Albacore* tick. As a fighting entity in the submarine force, she would never have been known without him. He was kind of a deep thinker and he was interested in what the *Albacore,* with its fantastic speed and maneuverability, could do to revolutionize naval warfare. He was out to prove that that submarine was the forerunner of the future for the submarine force."

The Boyes/Davis team shifted the emphasis from the scientific to the operational. Alongside the carefully planned series of tests that

The Wildman meets the Bellman—change of command ceremony, with Jon Boyes taking over from Ken Gummerson as Albacore's *CO. Gummerson compared himself to Lewis Carroll's Bellman, clueless skipper of the snark-hunting expedition. Jon Boyes called his ship's company "the wild men" referring to his no-holds-barred approach to radical tactics and maneuvering. (Courtesy of Ted Davis)*

yielded hard hydrodynamic data under the direction of the David Taylor Model Basin (DTMB), they began their own trial-and-error exploration of the *Albacore*'s performance, finding out exactly what she was capable of, and taking her head-to-head with the Navy's Antisubmarine Warfare Forces off Key West later in the year. As Captain Davis puts it: "Every minute the DTMB wasn't running a trial, we were running our own experimental tests, primarily aimed at tac-

Three Albacore *officers, including Lts. Ted Davis and Jim Ferrero, center, and right, visiting patrol plane squadron VF-82 to learn ASW tricks from the enemy himself, November 16, 1955. (Courtesy of Ted Davis)*

tics—and as a result the sub force learned a lot—though they'd never admit it!"

Admiral Boyes started with the conception that a submarine was organized around three teams or systems: the pilot or control team in the control room, the power team aft of the control room, and the fire control team in charge of weapons. But "this is not the way you fight wars," he said. Crew and systems had to be wedded to make a single team with power and control systems subordinated to, or rather adjuncts of, the attack system.

He was a member of the Naval Academy's first three-year class, class of '44. He spent the war in destroyers and then applied for subs. He found the Navy's small postwar submarine community (perhaps 1,100 officers and three times that many crew) to be a source of constant ferment and challenge. For example, he came to *Albacore* from command of the *Sarda*, then testing an elongated conning tower with advanced fire control systems. Later he commanded a squadron of nuclear boats and continued to experiment with hydrodynamics and automation. But he says when he was first advised that he was reassigned to *569,* he objected; he wanted to stay with active-duty, fleet-going subs and was "reluctant to be buried in research." But he went to Portsmouth, received a "warm cordial introduction" to *Albacore,* and went to sea with her. Almost immediately he realized what was possible if submarines were given enough speed and maneuverability, and realized he had been offered the most influential submarine of his time as his second command.

But what was the challenge that Jon Boyes, Ted Davis, and the crew of the *Albacore* faced? First, her submerged speed of 27 knots was much higher than the submerged speeds of any previous submarine; the Type XXI U-boats were usually rated at 17.5 knots, the *Guppies,* 15 knots (both on snorkel dieseling, not on battery!). In addition, *Albacore*'s short length offered an impressively tight turning circle, and her initial Phase I stern control surfaces offered instant, dramatic response at all speeds.

The Phase I stern, unlike those of all later single-screw submarines, mounted the rudders and diving planes aft of *Albacore*'s monstrous 11 foot prop. These movable planes were attached to fixed fins or outriggers that surrounded the prop. Later subs, including all U.S. nuclear boats from the *Skipjack* class on, mount rudders and stern dive planes on fixed fins before the single prop, much like the fins and control surfaces of the typical blimp. The difference is that the prop wash from the huge screw acted on the Phase I control surfaces to provide the ultimate in maneuverability at even the lowest speeds; she could turn on a dime at 4 knots or less.

This type of stern was used on the German Type XVIII U-boat powered by the Walter hydrogen peroxide turbine, the first serious alternative to electric power for submerged propulsion, as well as the Navy's first *John Holland* boat. According to Captain Jackson and Professor Allmendinger, it was used to replace an earlier arrangement of planes before the screw, which proved ineffective.

The Phase I stern gave *569* instant response; the teardrop hull gave her unprecedented speeds. And "single-stick control" gave the power to use this instant response. Previous subs had been controlled by a diving team consisting of a helmsman standing at a typical ship's wheel facing forward and two bow and stern planesmen facing the port

Albacore *in action! Only rarely did* 569 *do her stuff, as in this high-speed surfacing, where anyone could see it. Her radical high-speed maneuvering was usually 100% invisible, beneath the surface of the sea. (Courtesy of Russell Van Billiard)*

bulkhead of the control room and handling similar large wheels, all three under the supervision of the diving officer. *569*'s system gathered control of rudders and both sets of planes into an aircraft-style "yoke" or steering wheel, turned to control course, pushed forward or pulled back like a joystick to control dive and climb. In theory, *Albacore* was as much under the control of a single mind as is an aircraft due to its pilot's ability to respond and maneuver as quickly as he can think and use his controls, without the need for several individuals to respond to spoken orders and coordinate their actions.

The problem, of course, was that instant response gave the pilot instant ability to throw the submarine out of control, with possibly fatal results. Plainly, no one had any experience of the hydrodynamic forces acting on a submarine maneuvering at 30—or even 20—knots before *569,* and they were only partially understood.

Between *Holland* and *Albacore,* practically no hydrodynamic study of submerged submarines was conducted (except, perhaps, on the British "R" class killer subs of World War I). Speeds were so low that it was perhaps irrelevant. The Germans did conduct wind tunnel tests on the Type XXI U-boat and DTMB did towing tank tests on the *Guppies,* only to find their conventional hulls creating unexpected hydrodynamic effects. According to Capt. Frank Andrews—submarine project officer at DTMB from 1953 to 1954, later in command of the search for *Thresher* after her loss on diving trials in 1963, currently professor of engineering at Catholic University in Washington, D.C.—the *Guppy* boats proved very difficult to control in high-speed dives because of their large, flat deck areas. The angle of the dive would continue to increase in spite of reversal of the diving planes, as the deck itself began to act as a huge diving plane. Captain Andrews pointed out that *Albacore*'s symmetrical "body of revolution" hull form was the key not only to her speed but also to her three-dimensional hydrobatic maneuverability, giving her the essential quality of "inherent dynamic stability." *Albacore* had a minimum of surfaces creating unbalanced hydrodynamic forces tending to throw her out of control at high-speeds—essentially, only her sail. Her huge fins acted like feathers on an arrow, returning her to a straight and stable course when placed in a neutral position.

Nonetheless, some effects, particularly of the sail, were to prove worrisome, complex, and dangerous at *Albacore* speeds. Not only was it the major remaining source of resistance (Dr. Friedman says *Skipjack*'s large but streamlined sail contributed up to 30 percent of her total resistance), but it was also the source of the annoying and dangerous "snap roll." When turning, the sail acted as a wing, creating hydrodynamic lift. Since the sail is above the sub's center of gravity, this force would roll the submarine over toward the inside of the turn, robbing her of speed and of depth and course control, until she returned to an upright position. But if the loss of depth due to the snap roll is great enough, the submarine will be lost. Just as an aircraft out of control in a spin can attempt to recover until it has run out of altitude and crashes into the ground, a submarine diving out of control is lost when it passes its "collapse depth," that is, the point where the strength of its pressure hull is overcome by the pressure of the tons of seawater surrounding it, and it fractures and implodes.

But while an aircraft may (or may not) have thousands of feet in which to recover, the vertical dimension over which a submarine can operate is limited. World War II fleet boats were rated for an operational test depth of 400 feet, with a collapse depth of 600. The *Tang* class had an operating depth of 700 feet, with a collapse depth of 1,050 feet. The *Threshers* are thought to have operating depths of 1,300 feet. The titanium-hulled Soviet *Mike*-class submarine *Komsomolets* sank in the Norwegian Sea in April 1989 in waters more than 5,000 feet deep without actually collapsing. *Albacore* was repeatedly described as having a test or operating depth of 600 feet; and as her seventh commanding officer, Bill St. Lawrence, put it: "It was like flying a plane at 600 feet—one blink and you're in the ground."

A confidential memo from Admiral Grenfell, ComSubLant, to the CNO strongly urging *Albacore*'s decommissioning at the conclusion of her Phase IV trials (December 1962–March 1965) emphasized this problem: "Operating a submarine with a 600 ft. floor at 36 knots is comparable to flying an aircraft with a 5,000 ft. ceiling at Mach 2. It is extremely hazardous. That such is the price for progress is recognized. There is a proposal to provide *Albacore* with a second set of small control surfaces to be used during high-speed operations. These control surfaces will improve control, but they will not in themselves assure recovery should control be lost. Adequate testing of the new control surfaces will include the determination of the effects of extreme operating positions. Such tests involve considerable risks especially in a vehicle of this shallow test depth."

Eight years before, of course, the maneuvers and speeds Jon Boyes put her through had never been experienced in any other submarine. On the other hand, instead of viewing her test depth as shallow, *Albacore* was at that point one of the deepest-diving submarines in the world. She was the first to be fabricated from the new high-strength, low-carbon-steel later known as HY-80 (80,000-pounds-per-square-inch yield strength). Testing this type of steel was also one of the original purposes of her design. Although this steel proved the route to much greater diving depths, *569* could not fully exploit its strength. It was used only for her pressure hull itself; her framing was of conventional 50,000 psi high-tensile steel. Admiral Boyes says the welds were also a weak spot, unable to take the pressures the plates themselves could carry. The first nuclear subs to use it were those in the *Skipjack* class, also framed with the 50,000 psi steel. HY-80's fabrication proved extremely difficult; both they and *Dreadnought,* the first British nuclear boat, developed numerous hairline cracks in their hull welds. Actually, the *Threshers* were the first to be able to take full advantage of HY-80; the piping, and hull penetrations (periscopes, shafts, water inlets) of the *Skipjacks* could not handle the pressure.

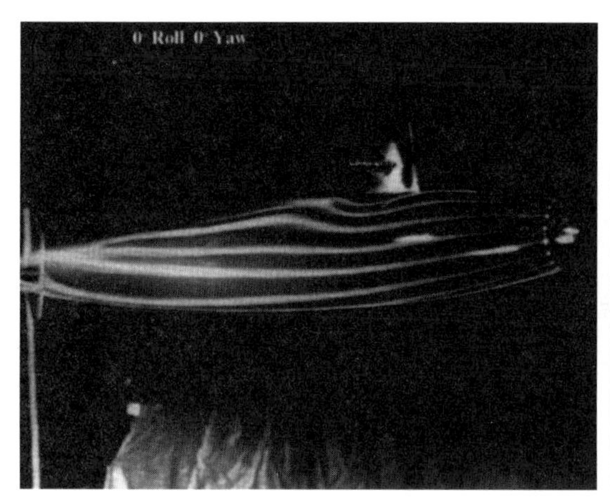

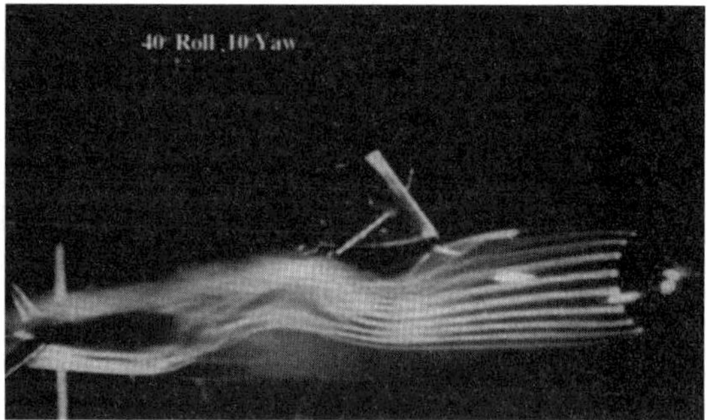

Physicist Dr. Henry Payne later worked on solving the "snap roll" with Jon Boyes. They conducted "smoke tunnel" tests of the Skipjack; *the plate to right shows the vortex created by the low pressure area next to the sail which pulls it over in a turn. (Courtesy Dr. Payne)*

However, Admiral Boyes tells of taking *Albacore* to depths much greater than 600 feet; he says 800 feet was a normal operating depth. On one occasion, they went to 1,400 feet in trials: "They called for us to go to maximum flight speed, then roll her over in a direct descent with a one-pilot override. He [the pilot] was mesmerized by the instrumentation and we were at 1,200 feet before we caught it." He says there was none of the creaking and groaning that might have signaled imminent danger. A later CO told him of an even deeper descent, when she had her "X-stern," a little slower in reaction than her Phase I configuration. She overflew her automated control, which was programmed to bring her out of the dive at 1,200 feet, and hit 1,600. "They never told the Pentagon or David Taylor," Boyes said.

Many things could trigger a fatal descent below collapse depth. Like anything else mechanical, diving planes could jam, most likely when moved all the way up to their stops, in a crash dive, for example. Stuck on full dive at high-speeds, this would mean disaster in a matter of seconds, a reason that so much effort was later spent on various forms of "dive brakes."

The difference between *Albacore* and her successors is that Jon Boyes (and some of her other skippers) repeatedly and intentionally took her to the limits of control to learn for the first time how submarines behave in such circumstances. Also her Phase I (and later Phase III X-stern) gave her the power to maneuver more radically than any other U.S. sub. According to Dr. Henry Payne, *Albacore* under Jon Boyes first encountered the snap roll—which submariners called the J.C. maneuver (J.C. for Jesus Christ!)—and later solved it. He and Jon Boyes participated in wind tunnel tests of *Skipjack*'s sail, including use of the "flying wind tunnel," the Navy's last blimp, carrying a model of the sub on a 33-foot-long retractable strut projecting from her belly radome.

Phase I: Maneuverability

Jon Boyes's concept of a single pilot "flying" *Albacore* led him to institute the practice of placing his executive officer in the pilot's seat at the controls, rather than the usual enlisted rating. Ted Davis became her most proficient pilot and describes some of the hydrodynamic effects in laymen's terms:

"The first control surfaces aboard *Albacore* were a simple set of bow planes in the normal old position, and a cruciform stern which placed large hydraulic stern planes and vertical rudders aft of the 'Big Wheel.' Control was instantaneous and dramatic. I was the designated pilot and my yeoman was the copilot. We seldom used anything but single-stick control. At maximum speed (30 plus knots) we would put about 10 degree dive on the stern planes for about 5 seconds and the ship would assume a 30 degree down angle. We did have rate-of-dive indicators which would tell us the rate at which we were headed down. That information wasn't really so critical or even important to us. What it did provide that was really important was the increase or decrease in rate-of-dive. When at 30 knots, 30 degree down angle, changing depth at 30 feet per second, it was nice to see the rate come off (decrease) when we pulled back on the stick. Sorta warm fuzzy feeling!"

Guppy skippers had already begun using such radical maneuvers as tactics for escape. Capt. Ned Beach, in his novel *Cold Is the Sea,* suggests that he took the *Amberjack* into 30 degree dives. This so-called dipsy doodling was unprecedented; the normal maximum angle for diving and surfacing was 2 degrees during World War II. Anything beyond 10 degrees, at that time, and the boat was assumed to be out of control.

Albacore was obviously doing things that had long been considered highly dangerous. Ted Davis says some crewmen found them unnerving; the solution in some cases was to give them the opportunity to handle the controls, thus making the motions of the ship intelligible.

Lt. William J. Herndon, LCDR Gummerson's executive officer, piloting Albacore *using her novel one-man "single-stick" controls. The wheel or "stick" was operated like an aircraft's, turned for left or right, moved backward or forward for climb and dive. This circular wheel and limited instrument panel are not those which she possesses today. (Naval Institute)*

They could understand then what she was doing and how she was (usually) well within the control of her pilot. "There were people aboard," Davis said, "particularly back aft in the electronic spaces, that were scared to death at some of the maneuvers and I would bring them up and say: 'You drive. Then you won't be afraid. You'll see how easy it is and how much fun it is and how you can dipsy doodle and still have

Albacore *was the first sub to experience the "snap roll." In a sharp, high-speed turn, her sail would act like a wing and roll her over. Her unique dorsal rudder at the trailing edge of her sail was used to counteract the heel. The one she carries now is larger than the original version. (Largess)*

The only other feature to detract from the perfect streamlining of her hull form besides her sail is this hump surmounted by a bit of flat deck which housed her diesel exhausts and mufflers. (Largess)

complete control and not be worried about getting into trouble.' They'd try it and go back aft and most would be fine from then on. But some would still be scared silly and so we'd no recourse but to transfer 'em."

The Dorsal Rudder, Keel Wing, and Bow Planes

Actually, the unbalancing forces on the sail had been anticipated by the *Albacore*'s designers, but the solution they provided was only partially satisfactory. This was *569*'s unique dorsal rudder, never tried on any other submarine. It could be extended out toward the inside of the turn, creating an opposing force to the lift generated by the fixed surface of the sail and canceling it out. It was worked by a pair of foot pedals, much like an aircraft's rudder pedals. But to extend this additional plane area out into *Albacore*'s "slipstream" created much additional drag, turbulence, and noise, as well as placing tremendous structural loads on the sail. It also added another complication to control. Captain Davis says that the normal procedure was simply not to use it and accept the heel:

"On a full-rudder, high-speed turn, the rudder at first acted like a sea anchor and radically slowed the ship," he explains. "To prevent the bow from going up, large amounts of stern planes were required to hold it down. Once the ship reached a constant speed, the stern planes could be zeroed and the heel angle would spiral the ship down. Incidentally, full rudder could be used in an emergency down angle to slow the ship and create an instantaneous squat. Just don't leave it on! There were a couple of crew members who could never get used to 30 degree down angles accompanied by 30 degree heel angles at high-speed."

Sudden reversal of controls or power at high-speeds to explore the parameters of her behavior produced heels of over 40 degrees and down angles of 50 degrees. To help the crew keep their feet, leather hand straps were installed—just like those you hold on to when you're forced to stand up on the bus or subway.

"*Albacore* could make a 360 degree turn faster than a jet airplane," said Davis. "But eventually too much speed will break down your maneuverability unless you can control it. *Albacore* found that high-speed turns sound great until you're hit with that squat and heel and lose all your speed. You might as well have thrown the damn anchor over."

The dorsal rudder was tried only on the *569;* the keel wing was never tried. Captain Davis says *569* performed better on stern planes alone: The bow planes were of little value except for exact depth keep-

Phase I Albacore *in dry dock. Note the presence of her bow planes, permanently removed during Phase II about one year later, as a noise-reduction measure. (PSM)*

ing near the surface. He says that in later submarines bow- and stern-planes planesmen typically spent all their time fighting each other, and he repeatedly demonstrated how much depth control was simplified and improved by shutting off the bow planes. *Albacore*'s were removed as a noise reduction measure in 1958: "We took the bow planes off the *Albacore* and our control problems went away. They were noisy, ill-placed, and served no purpose," he said.

Single-Stick Control and the Pilot Concept

Utilizing the *Albacore*'s speed and maneuverability to the full in actual tactics depended on the "pilot concept" and "single-stick control," which replaced the old slow-motion three-man diving team. Admiral Boyes, in fact, carried the concept to unique limits by putting his Executive Officer in the pilot seat.

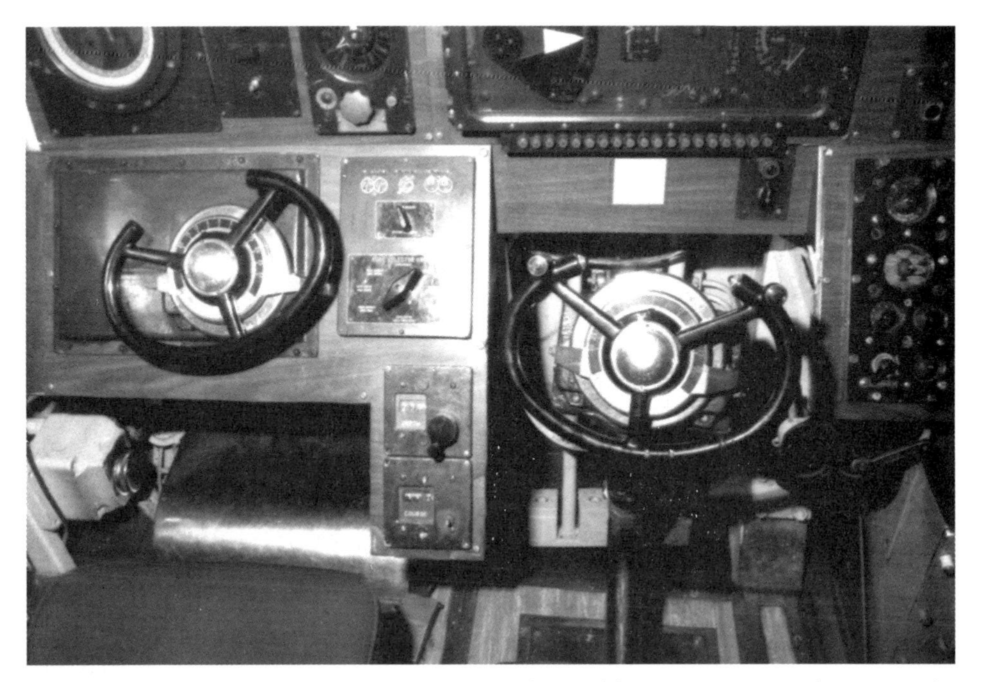

Albacore*'s present controls. This "yoke" can be used for one-man control; or control can be divided, with this yoke controlling the dive planes and the second wheel on the left controlling rudders. According to Harry Jackson, this yoke is taken directly from the Navy's large advanced blimps of the 1950s. Since the physics of blimp and submarine control are the same, he flew frequently on blimps to gain clues to* 569*'s behavior. (Largess)*

All nuclear submarines from *Nautilus* on have used two aircraft-style yokes similar to *Albacore*'s. One yoke is for rudder and bow planes, and one yoke is for stern planes, both of which were controlled by an enlisted man under orders of the diving officer—the same diving team as the fleet boats, just without the big stand-up wheels. All three functions can be gathered under the control of a single yoke, giving a single pilot and single-stick control. This arrangement, however, appears to have been seldom used.

On *Albacore*, though, Ted Davis says, ". . . Jon Boyes never said, 'Make your course such and such.' He would just say, 'Weapon on the starboard bow' or somebody closing from a certain position, and the guy who's the pilot maneuvered the submarine into the best position to evade or get a shot. This was the concept we were trying to use, and to this day the submarine force flatly refuses to entertain an idea like that."

This is the Combined Instrument Panel, with which Jon Boyes sought to make "single-stick control" a practical reality. It miniaturized the standard sub instruments and placed them in a panel where they could be quickly scanned. However, it still did not present the actual behavior of the submarine in a clearly visualized form. Note the row of lights at the bottom indicating the position of the dorsal rudder, controlled in this configuration by two thumb buttons on the upper rim of the "yoke." (Largess)

Davis, Boyes, and Payne all cite inertia and conservatism on the part of the submarine force to explain the failure to adopt the pilot concept, but another major impediment was Admiral Rickover, whose constant, overriding concern was the success and safety of nuclear power. His approach was to exclude other innovations from his submarines until they were thoroughly proved elsewhere. For example, the first generation of nuclear subs, *Nautilus, Seawolf,* and *Skate* classes, received conventional submarine hulls and were considerably slower than *569*. His emphasis was on safety and reliability over tactical qualities, ultimately producing a considerable reaction and the call for a return to the *Albacore* concepts by the late 1980s. This was fueled to some extent by the example of Soviet sub performance and design, which never really departed from the *Albacore* example. But Rickover opposed automation of sub controls. His philosophy was "Always keep a man in the loop," to

Barbel *was a diesel sub, the nearest thing to a direct though slightly enlarged copy of* Albacore. *This plan of her control room shows the need to reduce space requirements by as much use of automation as possible. Friedman says she introduced the modern control room arrangement, and push-button ballast control, tested on* 569. *(Naval Institute)*

counter the ever-present possibility of malfunctioning machines. Thus, the three-man diving team was ideal from his viewpoint.

The last thing Rickover wanted was high-speed maneuvering on his subs. Dr. Payne says, "He forbade it and panicked the submarine community—which still today refuses to deal with it." Payne credits Rickover with canceling the submarine wind tunnel studies for the very reason that they found the solution to the snap roll, making maneuvering safer and harder to oppose. The other side of the coin is that submarine tactics came to emphasize another characteristic—quietness—over speed and maneuverability. But this was in the future, not yet a concern with *569*'s Phase 1.

Implementing Single-Stick Control: The Combined Instrument Panel

One serious problem in implementing single-stick control was the lack of instrumentation. For Admiral Jon Boyes, reducing the number of people involved to speed up reaction time and reduce the possibility of human error or confusion was a major goal.

Albacore was delivered with the ordinary submarine instruments, depth gauge, inclination bubble, gyro repeater, bow- and stern- plane indicators. At this stage, flying the *Albacore* was only for the experienced. Ted Davis says, "It was thrilling—and it took good hand/eye coordination because of the lack of equipment to tell you what you were doing. You had to have a feel for it. My yeoman-assistant copilot was an expert; you could trust him to fly that thing anytime, any place."

Albacore had no combined instrumentation when she went head-to-head with the Navy's ASW forces in the fall of 1955. "All we had was experience," Davis remembers. "You just watched the depth gauge and rate of turn indicators," crucial because, "you could die of fright before you saw something happen," and these would show that the *rate* of dive, say, was falling off well before the submarine actually began to come out of the dive.

This characteristic—the time lag in response to control—was shared with blimps and is the reason that Jon Boyes and his officers went to Lakehurst to fly blimps and handle their controls, early in his command. *Albacore*'s original control yoke, control circuitry, and autopilot have been described as taken from, or based on, those of a blimp.

Automated Control

When Boyes arrived, *Albacore* had an "analog servomechanism," essentially a blimp autopilot. The problem was programming it; the characteristic behavior of a submarine traveling at *Albacore* speeds had never been encountered before. *569*'s first job was to produce it! DTMB provided the initial calculations for a maneuver; *Albacore* would go out to sea and try it, and DTMB would refine its calculations. Automated control eliminated human failures due to lack of coordination or slowness of reaction, pilot stress, fatigue, boredom, and distraction. *Albacore* was "steady as a rock" and much quieter, with less control surface motion "banging those planes back and forth," according to Boyes.

Tactics

Admiral Boyes writes that the ultimate proof of the pilot concept, single-stick and fully automated control, as well as speed and maneuverability, came during exercises against ASW forces, "particularly the tracking and attack of high-speed destroyers. *Albacore* was brought smartly to

Phase I Albacore *docking, showing retracted bow planes (later removed) and her second sonar, a troublesome and primitive JT, under the bow dome. Its purpose was to avoid collisions while surfacing. Study of the hydrodynamic and acoustic effects of the dome by DTMB scientists yielded great advances in sonar performance. (Naval Institute)*

periscope depth, the target was locked in, firing was simulated, and then *Albacore* was spun on her tail to go after other destroyers.

"If conditions were considered just right, *Albacore* was moved into an optimum position relative to the target. In such a situation, while holding the speed and maneuverability advantage *Albacore* could fire at very close range with low relative bearing changes—or might scoot under the target releasing simulated vertically launched missiles at very short range."

However, fully automated control and the CIP were still not available when *Albacore* was ordered from Portsmouth to Key West at the end of October 1955, for her first astonishing day of ASW trials, November 4, 1955. After arriving, Boyes describes meeting immediately with the aviators and destroyer people to plan the exercises, going without sleep for 48 hours.

For at 11:50 on the morning of the 4th, a Navy transport landed

When CNO Admiral Arleigh Burke invited Lord Mountbatten to ride the Nautilus, *Rickover vetoed it. Burke took Mountbatten to ride* Albacore *at Key West on November 4, 1955, as a consolation prize. (National Archives)*

at the Boca Chica Naval Air Station bearing Adm. Arleigh Burke, CNO, and Adm. the Earl Louis Mountbatten of Burma, First Sea Lord. Met by an assortment of high-ranking officers, including Rear Adm. Frank Watkins, ComSubLant, they then boarded a car for the Key West Naval Station and were driven to Pier Baker, where *Albacore* was moored. Minutes later, under a typical blue Key West sky, 80 degrees, smooth sea, *Albacore* was underway with them aboard.

Mountbatten, newly appointed First Sea Lord, was faced with challenges from three sources: the heavy burden on the Royal Navy (RN) as sustainer of Britain's vast peacetime commitments, the need to bring the RN into the nuclear age, and the determination of the Exchequer to cut and reduce the RN to the position of the least of the three services. *Nautilus* had been launched in 1954 and Mountbatten immediately saw the nuclear submarine as the means to revitalize the RN and establish its significance for the foreseeable future. Personally close to Burke, he came to the United States in October 1955 to woo with his support the opponents to transferring nuclear information in the Atomic Energy

According to Capt. Ted Davis, then Jon Boyes' XO, it was this day with Burke and Mountbatten aboard that Albacore *achieved her first submerged speed record* and *went head-to-head with the Navy's ASW forces for the first time. (National Archives)*

Commission, the Department of Defense, and the U.S. Congress, something possibly forbidden by law. Rickover was also opposed to sharing such information with anyone, allies or not. When he learned that Burke had promised Mountbatten a tour, or maybe the chance to go to sea on the *Nautilus,* Rickover refused to allow him aboard, even though he had been showing her off to every possible dignitary and politician he could find since her commissioning. Arranging the trip aboard the *Albacore* was the best Burke could do for a consolation prize.

Within a year, however, Mountbatten had won over Rickover, paving the way for the installation in 1959 of a *Skipjack*-type reactor in *Dreadnought,* Britain's first nuclear submarine. As last viceroy of India, Mountbatten has sometimes been faulted for failing to make

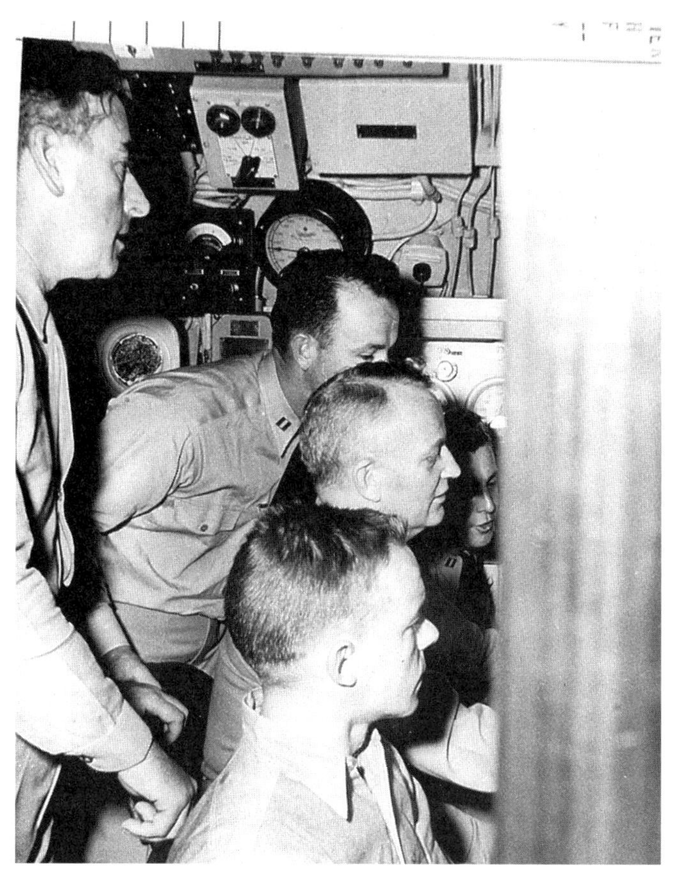

Admiral Burke at the "stick." Both he and Mountbatten took a turn at the controls to "fly" the Albacore, *discovering with delight how simple single-stick control made it. (National Archives)*

peace between Hindus and Moslems. But if he could charm Rickover after that slight, his powers as a diplomat must have been superhuman. As one of his staff said: "Rickover didn't give a damn whether we as a country got the submarine or not, but he did care whether Lord Mountbatten got one!"

As Boyes describes it, they came aboard leaving politicians and brass on the dock; Burke had intentionally left everyone else behind. He and Mountbatten greeted Boyes warmly, saying, "We know this is a bit of a strain, but relax, we're your friends," and *Albacore* left the dock with the three of them packed on her little bridge.

First they headed for the measured two-mile course off Key West, to try to break, according to Admiral Boyes, the submerged speed

In the end, it was the Soviet Navy which made the effort to develop to the fullest the characteristics of Phase I Albacore. *- high submerged speed and maneuverability, where the US came to emphasize quietness and sonar. This Soviet* Victor *class SSN in 1974 shows her stubby, bulbous* Albacore-*hull. She goes* 569 *one better, replacing the sail with a low, rounded oval to eliminate the snap roll. On the other hand, her broad, flat deck detracts from her streamlining. (NHC)*

record of 30 knots set by the new hydrogen-peroxide-powered British sub *Explorer*. Mountbatten said he "hoped they could best them." Here, Boyes says, actually over the last 1 1/4 miles of the run, they set the record that lasted till *Skipjack* came into service.

Then they headed for the ASW exercise zone. "I sent them below and we dove," said Boyes. "I was bursting with pride—the ship was sparkling, and there never was a finer crew. But the professional pride of the patrol planes, blimps, and destroyers we were going up against was at stake too; so they had a spotter plane to the west to pinpoint us before we submerged. We headed due southeast at maximum speed, then cut back up. The sonarmen picked up the *Sarsfeld* and we headed for her blind spot—bow on. We went straight for her, closed her at high-speed, and fired 'green flares' against her. Then we went for the other destroyer and did the same. Then we went deep, skirted the blimps' and VP's MAD and sound buoy barrier, and arrived at the Sea Buoy ahead of schedule, returning to port undetected. Mountbatten came into the wardroom for coffee and said, " 'That's the greatest damned show I ever saw in my life!' "

On the return to Key West, Burke and Mountbatten each took a turn at the controls with delight. Ted Davis recalls: "We were approaching Key West in shallow water and I finally had to reach out and pull back on the stick and tell them, sorry, that's all for today."

Afterward, *Albacore* continued to develop her tactics against the Navy's ASW forces. Boyes would concentrate on the fire control problem, leaving the tactical moves up to Davis. The first thing they would try was to go for the bow position. The speed of the sub and destroyer combined added up to a 40- or 50- knot relative speed, far higher than the 20 knots that was the best at which the escorts could get a sonar indication. If they tried to launch weapons, they'd fall behind the sub. Their second choice was to come in very deep amidships, where the DDs were equally helpless. They handled aircraft by going very deep, with MAD giving detections to 200, or at best sometimes to 400, feet.

The *Gearing*-class destroyers of the time were barely a match for *Guppies*. Their best weapon was the "Hedgehog," which fired a small cluster of rocket-propelled bombs in a shotgun pattern over the bow of the ship. And *Albacore* could simply dive away under the Hedgehogs, faster than they could sink. But with her tight turning circle, no destroyer could bring them to bear in the first place, or even hold a contact on her, or operate their sonars at all the speeds necessary to chase her. Another trick she used was to change course at high-speed, creating big "knuckles," or clusters of bubbles. Sonar impulses would bounce off the knuckles and contact would be lost.

Submarine versus Submarine

Captain Davis has continued to argue for the need for "dogfighting" maneuverability in sub-versus sub-combat. He had the rare experience of dogfighting with a Soviet conventional sub in his command, the *Grenadier,* a "Guppy" conversion, in May 1959, sticking on her tail till she ran out of air and battery and was forced to surface. He says: "Predators depend on speed and maneuverability one minute, stealth and cunning the next; two items on a survival menu. . . . In the Navy's enthusiasm for high-speed deep-diving submarines, it overlooked low-speed maneuverability."

But this is the quality the Phase I stern gave *Albacore*: "You could turn that submarine on a dime," said Davis, "put that rudder over, and then jam that power on for 10 seconds. It'll wheel that sucker right around and, boom, you've changed course 50 or 100 degrees at a moment's notice and then put that rudder amidships and, zoom, you're gone.

"Now if you take the present-day nuclear sub, put the rudder over, and jam on full power, nothing is going to happen, absolutely nothing until you get up to 6 or 8 knots, and then you'll begin to turn.

"Say you're in a dogfight, you've got a contact on your quarter. You better do something or you're a dead cookie. *Albacore* only needed four or five screw turns to turn her nose on him and get off a snap shot. But to turn around with her Phase II stern took a large circle and lots of noise."

On the other hand, *Albacore*'s huge rudders and stern planes aft of the prop were arguably dangerous—too much control surface at higher speeds. It was far too easy to throw her out of control through a minor pilot error. A solution was the installation of aircraft-style "trim tabs," that is, very small control surfaces adjacent to the big ones, which were controlled by a small wheel located beside the pilot's seat. Besides compensating for instabilities while using the big planes, it later turned out that the large planes were not really needed at all, except at very low speeds of 5 knots or less. At all other times the trim tabs alone were safer. Ted Davis says: "They came in handy for long-distance, high-speed transits; we'd get them set perfectly and avoid flapping those big planes around."

Other Experiments, Other Problems

Jon Boyes's concept of flying a submarine like an aircraft also involved reducing the number of personnel aboard, as well as putting her under the control of a single pilot. This was in opposition to Rickover and the submarine community, which favored large crews for safety and maintenance. "Our crazy idea," said Boyes, "and you were a scoundrel for even talking about it, was a control room with a CO, pilot, and a crewman for flooding and blowing tanks, 3 men versus the 10 normally needed." He managed to halve the number of men needed to run *569*. They experimented with a Control Engineer Panel, allowing one crewman to operate the blow and vent, and trim manifolds. "There were too many people pulling levers," he said. "The idea was . . . instead of having the Chief of the Boat mechanically pulling valves, let a solenoid do it."

Albacore also pioneered a single, multipurpose mast for a number of antennas. This idea, incorporated in the BRA-34 masts used on the *Los Angeles*–class submarines of today, permitted a reduction in sail size. *Albacore* also carried a single experimental periscope that combined the functions of the usual thin, inconspicuous attack scope and large-aperture search scope.

Lacking torpedo tubes, Boyes used the emergency signal ejectors to launch green flares simulating missiles when he bored into an attacking position on a destroyer. Thinking about the possibilities, they approached Aerojet; engineer Cal Gonwer designed a working missile for them, a very early precursor of the submarine-launched Tomahawk missiles used to strike precision targets in Baghdad during Desert Storm.

Not everything on *569* was perfect, however, according to Ted Davis: "The ship was delivered with a JT sonar (rudimentary) from an old submarine. We had to have something that would provide us with an 'all clear' when we surfaced. It leaked, leaked, and leaked. More hours, hate, and discontent were devoted to that damn thing than all other troubles combined." It was originally a rotating T-shape antenna projecting from the deck before the sail, as in World War II boats. Later it was mounted under a dome between the bow planes. Admiral Boyes says New London took off the old JT head and completely redesigned it. Afterward, trained operators got remarkable performances from it: "Scott could hear through our own white noise at 30 knots; you could filter that out if given enough time." *Albacore* found she could track *Guppies* at great distances because her streamlining made her so quiet.

End of Phase I

In December 1955, *Albacore* entered dry dock at Portsmouth Naval Shipyard to undergo her first stern replacement. With her original stern went *Albacore*'s incredible low-speed maneuverability, together with perhaps the most exciting period in her history, a period that provoked controversy for years to come. Phase I demonstrated her unique and unfulfilled promise, a submarine piloted and flown like an aircraft, performing "hydrobatics," maneuvering freely in three dimensions beneath the sea. Yet these achievements were largely forgotten when, with Phase II, the Navy began to explore other directions of submarine development.

Nuclear power revolutionized the capabilities of the submarine, giving it complete independence of the atmosphere and unlimited submerged endurance and greatly increased submerged speed. Yet in 1955 the Nautilus *was in many respects a conventional boat. Totally committed to the success of nuclear power, Adm. Rickover was unwilling to burden her with further innovations. (US Naval Institute)*

V Phase II: The Fifth Revolution

HARRY JACKSON HAS WRITTEN OF FIVE REVOLUTIONS in the history of the submarine: The first four are the *Holland,* the first practical submarine; the adoption of the diesel, giving long range and endurance; the fleet boat, perfection of the "submersible"; and the *Tang,* the first U.S. boat designed for submerged performance. The fifth revolution is the marriage of the *Albacore* hull and nuclear power. This came in 1959 with the *Skipjack,* the seventh nuclear sub, combining the independence of the atmosphere and (relatively) unlimited power of the nuclear reactor with the form, speed, and maneuverability of the true submarine.

The *Nautilus* "underway on nuclear power" on January 17, 1955, had an astounding impact on popular—and naval—opinion. Other than her speed, nobody really understood what the peculiar little *Albacore* was doing. But the *Nautilus* seemed like the fulfillment of Jules Verne's dream of a spaceship under the sea, able to voyage freely in an unknown world. Her success required an even more revolutionary engineering effort. It was early considered building her as an unarmed test vehicle, but Admiral Rickover foresaw that that would lessen her impact on the Navy. It was essential that nuclear power be introduced in a fully operational man-of-war, like the *Monitor* or the *Dreadnought,* making the world's navies obsolete at a stroke.

Another proposal was to give her an *Albacore* hull. But Admiral Rickover was determined that *Nautilus* be subjected to no innovations *except* nuclear power. Norman Friedman notes that her narrow, tapered stern was designed to take the single prop indicated by Series

Here is Albacore *in her Phase II configuration, with fins with large fixed areas and control planes ahead of the prop. This became the prototype stern for all U.S. nuclear boats from* Skipjack *on. (PSM)*

58. But Admiral Rickover insisted she be given twin screws for the sake of reliability. For safety's sake, she carried a complete diesel-electric propulsion system with batteries and snorkel.

Hyman Rickover has often been criticized as a conservative force in the submarine program, holding back progress in all areas except nuclear power itself. Later, he gave much attention to other factors such as speed, quietness, and diving depth, but only after the safety and reliability of nuclear power was proved.

One unexpected result was the near total control he took over the nuclear sub program, personally selecting all officers after subjecting them to grueling, even humiliating interviews, and requiring them to accept his "discipline of technology," with a perfect engineering mastery of nuclear power, as well as the skills of a submariner. Officers, crews, design teams, all had to meet truly ferocious standards of education and competence. And it worked; the U.S. Navy never had a nuclear accident.

This is in contrast to the Soviet navy, which lost a number of nuclear boats to reactor fires and radiation contamination. Perhaps this was the result of a conscious decision to cut corners and take daring technological risks in their determination to catch up and surpass the Americans.

But guess whose job it was to take those risks in the U.S. Navy? If *Nautilus* and the nukes had to be flawlessly reliable, it was the

Albacore's job to be the Snark—to test out everything unknown and unpredictable.

Rickover's willingness to explore new technology but to decide absolutely in favor of the practical and reliable is shown in the second submarine reactor, the *Seawolf*'s liquid-sodium cooled plant. It was smaller and highly efficient, but when it proved dangerous, complex, and full of defects, he dropped it forever in favor of the big, bulky, noisy, but inherently simple, safe, cast-iron reliable pressurized water reactor. Rickover proved wiser than the submarine community in another respect; he very early foresaw the value of high-speed, and of very large submarines. Few remember that the *Nautilus* was a huge submarine for her time. At 323 feet long and 3,533 tons, she was exceeded only by the Japanese I-400-class aircraft carrier submarines, far larger than any attack submarine ever built. This flew in the face of submarine tactical thinking; Rickover was forced to accept a class of smaller, more maneuverable *Tang* equivalents, the *Skate*s, to get the very large, fast submarine he wanted. This materialized as the giant radar picket submarine *Triton,* with twin reactors.

Mighty *Triton* broke 30 knots (on the surface! She was designed to keep up with fast carrier forces). But she had already been beaten for the record of fastest submarine in the world by the *Skipjack*—submerged. Brute force engine power had been beaten by the simple, elegant solution of reducing resistance—with the *Albacore* hull.

Still, big, noisy *Nautilus* should have been an easy mark for sonar, but her speed made her an insuperable opponent, as she went head-on with the U.S. and British fleets and very nearly "sank 'em all." She closed ship after ship in exercises at 24 knots "attacking almost at will," as Norman Friedman describes it. Then with a burst of speed she would make her escape; the short-ranged destroyer sonars of the day could not hold a contact on her. On a single day she popped up in positions a hundred miles apart, or hid undetected under an "enemy" carrier in the center of the task force for hours on end. She outran homing torpedoes; she was too agile to be caught with nuclear depth charges. Once she revealed her position with a flare and was jumped by a screening helicopter, but then she was already 2 miles away, out of range of any weapon. Pretty Snarkish behavior indeed, for the great big "showboat" with her three decks, wood-grain paneling in the wardroom, and the only staircase ever carried by a submarine!

Moviegoers, of course, will not be surprised, remembering how "Jaws," in spite of weighing multiple tons and being 30-some feet long,

Another representation of Phase II Albacore, *a model on display at the Mariner's Museum in Newport News, Virginia. The Phase II stern reduced drag and reportedly gave greater stability at high speeds at the price of some of Phase I's fantastic low speed maneuverability. (NHC)*

was quite good at sneaking up on folks and gobbling them down in a flash. Not a bad portrayal of the job of a modern nuclear sub: to be there, invisible, hang around as long as it takes, then eat you up when you're not looking. The *Nautilus* was, shall we say, the Navy's first great white shark. But she could have been defeated within a few more years without *Albacore* to teach her successors some Snarkish tricks.

Phase II: The Skipjack Configuration

Series 58 was only the beginning of David Taylor's hydrodynamic studies of *569*. Series 58 essentially only measured resistance; the forces acting on her in maneuver were more complex. In 1956, two fiberglass "tethered" models were constructed, powered by a cable connected to an overhead carriage that followed the model around the tank. Under the direction of Morton Gertler, these were used in studies of stability and control. Predictions based on the model tests were very close to the actual behavior of *Albacore* in full-scale trials, according to Jerry

A rare photo of Phase II Albacore, *this view of her in another sub's periscope shows some Phase I features, the sonar dome and bow planes, which she lost later in Phase II. (NHC)*

Feldman, head of DTMB's Submarine Dynamics Branch, and Bill Stensen, of its Full Scale Trials Branch.

Series 58 had tested a number of stern configurations, "+" and "X" planes, and twin contra-rotating props. There was some thought of giving *569* variable shape control surfaces, but the approach of simply removing and replacing her stern was adopted. Thus, *Albacore* was cut up and put back together three times:

Phase II—December 1955 to March 1956—giving her the *Skipjack* stern with control surfaces before the prop
Phase III—November 1960 to August 1961—the X-stern
Phase IV—December 1962 to March 1965—contra-rotating props, second main motor, and silver-zinc battery

Each configuration was tested beforehand with the tethered model. Then *Albacore* conducted full-scale trials, testing out the predictions, refining the theory. Dr. Gertler looked forward to a Phase V with the sail removed for perhaps a 30 percent reduction of drag and elimination of the snap roll. But this was never done.

Albacore's Phase II configuration was the prototype of the *Skipjack*. (In 1955, Capt. John McQuilkin and Cmdr. Edward Arenzen of BuShips Preliminary Design Branch had determined she would have an *Albacore* hull; Electric Boat engineers John Leonard, Russell Brown, and Harlan Turner designed her so. *Albacore*'s HY-80 steel now proved absolutely essential; ordinary high-tensile steel simply could not provide the necessary strength for a hull of such wide diameter.) When she went to sea on her trials, she took away *Albacore*'s speed record, probably breaking 30 knots (although a quieter prop later kept her at 29). But when Admiral Rickover's biographer Francis Duncan says, "Beneath the surface, the *Skipjack* had behaved like an airplane, banking and rolling as she maneuvered at high-speed," it must be remembered that in this respect, she was a step backward from the *Albacore*'s fantastic Phase I maneuverability. And perhaps that lauded banking was the "snap roll" and not such a fine thing?

Ted Davis regretted the abandonment of the Phase I stern and argued for its virtues; but its potential for overcontrol was great. It appears the Navy made a conscious decision for speed and stability over extreme maneuverability. The dorsal rudder was removed in Phase II also; whether as unnecessary (though the snap roll was still quite possible) or as an undesirable complication is unclear.

"If you interviewed ten skippers today," says Davis, "they'd say we don't need any more maneuverability, we never use more than 2 degrees or 3 degrees of rudder out of a possible 30 degrees available but they don't stop and then say, But we always travel at 18 knots—we never get to dogfight where you need maneuverability at 5 knots, 3 knots.

"They think weapons should do it—shoot a smart weapon a million miles away. But what I worry about is suppose all of a sudden they find this guy sitting right beside them—what tactics are they going to use at that point?"

According to Bill Stensen and Jerry Feldman, the Phase I stern was "very draggy" and Phase II was "probably the one they really wanted in the first place." It appears on the 30- x 6- foot fiberglass model tested in the Langley wind tunnel. The large fixed surfaces now added much stability—she now tended to level off and straighten out with the planes at 0 degrees.

However, Harry Jackson says there was much doubt as to its effectiveness, so "initial sea trials were commenced by hauling her into the stream by tugs, and when pointed toward the open sea applying power," before finding out whether they worked.

This painting by P. Melville shows the Albacore *and her contemporary, the lovely, graceful P6M Seamaster, the Navy's last, and only turbo-jet, flying boat. They have this in common: both forms derive from the extensive research by Stevens Institute. Stevens tested Navy flying boat designs for their water-borne performance characteristics in World War II, building a great fund of expertise which went into the Seamaster. (NHC)*

Jon Boyes's second executive officer (XO) and "pilot" for the Phase II *569,* Lou Urbanczyk, feels the Phase II stern was as maneuverable as the previous one. He thinks one advantage was to reduce noise by taking the planes out of the prop wash and doubts whether removal of the dorsal rudder indicates any reduced tendency to snap roll.

Phase II trials were from April to November 1956. Urbanczyk says *569* had just had a new pilot's seat and controls taken from a blimp installed by the crew and shipyard. He notes that when CO of the nuclear sub *Jack,* she had the ability to switch to "single-stick control" but never used it. He gives a whole new list of flaws in the pancake diesels: "weak retainers for piston rods . . . the spherical thrust bearing at the bottom of the engine wiped . . . the expansion joint for the exhaust kept breaking, not flexible enough." He wrote his command thesis on the pancakes; his conclusion: "They should get rid of 'em."

Vice Adm. Lando Zech was her third skipper, relieving Jon Boyes on January 15, 1957. Admiral Zech went on to command the *Nautilus,* the only *Albacore* CO to make the transition to nuclear power. He describes inducing the snap roll for DTMB: "They wanted to try this out if going at max submerged speed you threw the rudder full on. The sail is a much more controlling factor than the planes—you'll get a big down angle as well as heel at the same time. DTMB wanted to do it; I didn't. You had to have a feel for the angling and dangling of the ship; nobody knew for sure what would happen—would the ship right herself?

"I said, I'll go to max speed, put the rudder at 10 degrees and leave it on. We had good depth above and below. I asked 'Are you ready?' They thought I was overly concerned. But I expected a big angle, had everybody holding onto the straps.

"In seconds the ship took a 42 degree heel and 41 degree down angle. We lost 200 feet of depth; the breadboard instrumentation we had in the control room was falling all over the place. I rang up all stop, all back emergency, let go of the controls—the ship just settled down.

"Then I said, 'Now what do you want to do?'

"'Captain, that's enough, let's go back to port.'"

From March to June 1957, she went to Key West to work with destroyers evaluating their weapons. Twelve years later, the shock of the Type XXI was over. The best and newest World War II DDs, the *Gearing*s, had better weaponry than the Hedgehogs that the Phase I *Albacore* had gone up against. One *Gearing* was considered an even match for a *Guppy* and two certain to get it. They had over-the-side homing torpedoes improved from 12 knots speed in 1950, to 15 knots in 1955, to the Mark 43 with 30 knots and 6,000 yards' range in 1957. They had Weapon Alfa, a trainable launcher firing a rocket-propelled depth charge weighing 525 pounds with a range of 975 yards', versus 265 yards for the 20 pound Hedgehog. Their weakness was the short range of their sonars, requiring a tight, close screen.

"It was an interesting period, working with the destroyers," Zech remembers. "We'd go out on Monday—she was hard to maneuver on the surface, so slow—it was more convenient to stay out through Friday. We were on the surface most of the time. She was like a cork! If a big storm came up, you could dive, but then you'd be using up your battery. The inclination was to head into the sea and ride it out. You respected the sea—and picked your course to avoid the storms."

Both Zech and the DD skippers wanted seriously to win. "When we first arrived we had a meeting and the two COs were asked to share their ships' characteristics with each other. I was suspicious the DD skipper might not be so forthcoming," says Zech, "but I'd spent time on

On the other hand: blimp and submarine have much more in common. From the standpoint of resistance and control, they are essentially the same vehicle, operating in different media. At many stages of her career, Albacore's *designers and operators went to the Navy's blimp force to study their behavior and characteristics, borrowing specific items of equipment such as controls and autopilot. (NHC)*

*Gearing*s and had a pretty good idea of their speed, turning radius, acceleration and deceleration, etc. I went first and told him *Albacore*'s max speed, turning radius, max acceleration, turning angles, rates of ascent and descent. He was amazed. When it was his turn he said NO! I kind of felt this might happen.

"On the first exercise, I was told to stay in a known spot at a very low speed, shallow depth, set course, and I couldn't evade until I heard the weapon hit the water, like I was on unalert status! I didn't like it—not realistic—but I agreed, I was so confident *Albacore* could outmaneuver anything they threw at us. I gave orders, when we heard something to take max speed, dive at a big angle, go right or left. We outturned and outmaneuvered the torpedoes—we weren't necessarily faster. At the end of the exercise we were asked to make no report—the officer in charge and the DD skipper would do it—maybe say it was favorable? So I made my own report, sent it up the chain of com-

Harry Jackson has called the marriage of nuclear power and the Albacore *hull the "fifth revolution" in submarine capability. This took place with the* Skipjack *SSN-585. Yet she and her 5 sisters were the only nuclear boats to use the pure teardrop hull, as shown by the upper model. They were the fastest and most maneuverable nuclear boats for many years, but almost immediately the emphasis shifted to quietness and sonar. The* Thresher, *SSN-593, lost the pure teardrop hull to the need for enlarged machinery spaces aft to accommodate a bulky, "raft-mounted" propulsion plant, sound-isolated from the hull. (US Naval Institute)*

mand. I was called before the Chief of Staff, he told me I had not exercised proper protocol—but with a smile on his face; he wasn't all that disturbed."

Unfortunately for the subs, this cut-and-thrust infighting could not continue. The answer for the surface ships was much-longer-ranged sonars. At this point it began to look like being detected meant being sunk, automatically, and the emphasis shifted back to stealth—in the form of quietness. Not only does the submarine's radiated noise increase with speed but her ability to use her own sonar drops off quickly as well. The nuclear submarine with her noisy reduction gears for her steam turbines and water circulation pumps for her reactors is at a special disadvantage.

Albacore was the test bed for these advances, spending much of the rest of Phase II until November 1960 on the study of submarine self-noise, quieting techniques, and sonars. Today numerous examples of rubber mountings isolating machinery and piping from the hull are visible inside *569*. According to Howell Russell, DTMB scientist, "She was constructed without any concern for noise and went from being the Navy's noisiest submarine to the quietest."

Lt. Cmdr. Robert D. Thompson relieved Lando Zech on January 15, 1958. *569* was overhauled, then assigned to Submarine Development Group Two, a special unit within the Atlantic sub force commanded by Frank Andrews. Including *Tullibee* and *Thresher,* it was set up (Marvin Lasky says Captain Andrews was the driving force behind this) to develop technology and tactics for the use of submarines in ASW, as well as to familiarize research scientists from many institutions with the realities of submarine warfare. (Jon Boyes says *569* went up against *Guppies* very successfully because she was so quiet—streamlining drastically reduced flow noise. It was the internal noise that was the problem.) On completion of her overhaul in June, testing resumed, with emphasis on noise studies. In addition to resilient mountings, a water-based plastic Aquaplas was used to coat the interior surfaces of all free-flooding areas such as ballast tanks and superstructure, absorbing vibration and dampening water flow noise. (Flutter of the thin outer double hull was a problem.) One-half to three-quarters of an inch thick, Aquaplas did not bond well to metal; Portsmouth Naval Shipyard engineer Howell Russell remembers entering the ballast tanks and finding it coming off in big sheets. It has since been replaced by other materials. Hydrodynamic noise was reduced by streamlining measures, such as cleats that flipped over into

Here is the Barbel *in dry-dock. The three* Barbels *were the US Navy's last diesel subs, built in response to the often-made suggestion that* Albacore *be duplicated as an attack submarine. Cramming in the necessary three in-line diesels, torpedo tubes, fire control equipment, snorkel, was not easy and cost some speed. Apparently quite successful, these plans were supplied to Holland and Japan which built near duplicates. (US Naval Institute)*

the deck, hatches flush with the outer hull, and the clamshell covers over the flying bridge. In October 1958 the noisy bow planes and their hydraulic cylinders were removed. In 1959, she got a new 14-foot prop; this reduced noise by providing the same thrust at lower RPMs, thereby reducing cavitation.

She went to the "Tongue-of-the-Ocean" in the British West Indies for two weeks in May 1959, then went to Key West to work with the Surface ASW Detachment again. On July 4, 1959, Lt. Cmdr. William C. Rae took over, and in January 1960 she began a series of varied trials for DTMB, including evaluation of a highly unusual concave sonar dome. On August 24, 1960, Lt. Cmdr. Wallace A. Greene relieved Bill Rae as CO, and on November 21, she entered Portsmouth Naval Shipyard for her "Phase III" conversion. If her Phase II behavior was, indeed, a little more dignified, her full Snarkishness was about to return with a vengeance.

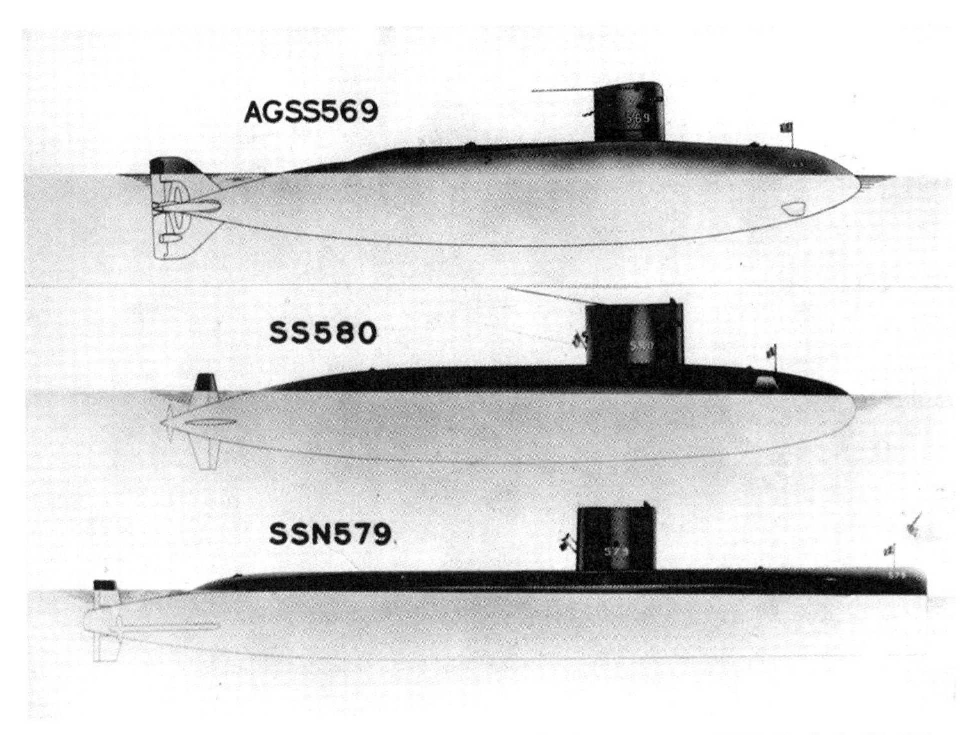

Comparing the Albacore *with two successors, built in 1958 at PNS.* Barbel, *SS-580, was the only combatant diesel boat with an* Albacore *hull.* Swordfish, *SSN-579, was the last nuclear sub class built with a conventional, non-teardrop hull.*

Diesel Postscript: The 3 B's

Most nuclear boats from *Skipjack* on are children of the marriage of *Albacore* and *Nautilus,* but *569* had three more out of wedlock. These are the *Barbel, Bonefish,* and *Blueback,* fulfilling numerous suggestions that *Albacore* should be directly copied as an armed and operational sub. The Navy's last diesel boats, they are externally similar to the Phase II *Albacore,* except for bow planes moved to the sail and a larger flat superstructure deck. Laid down in 1956 and 1957, all were commissioned in 1959. Their major difference besides torpedo tubes is the need for a more practical balance of diesel to electric power: three diesels with 4,800 hp, two motors with 3,125 hp (less than half *569*'s 7,500), so their submerged speed is a "mere" 18.5 knots.

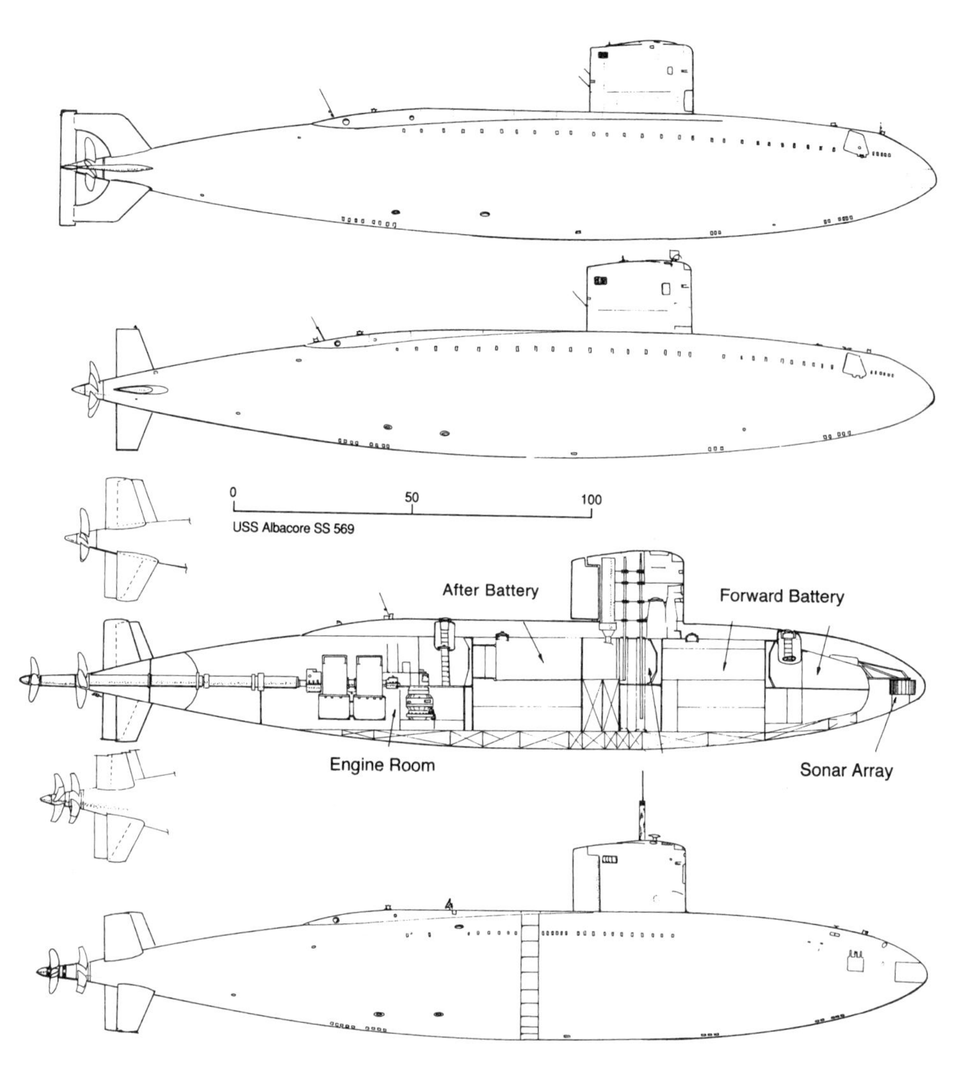

These drawings, by Jim Christley, show Albacore's *four stern configurations, Phase I with control surfaces aft of the prop, Phase II with surfaces before the prop, adopted for* Skipjack, *the Phase III X-stern and Phase IV contra-rotating props. All of these were studied in the 1949 Series 58 model tests at David Taylor. Jim Christley notes some details may be inaccurate. For example, the scrap view showing closely spaced contraprops represents a modification which was planned but never made. (From* U.S. Submarines Since 1945, *Naval Institute Press)*

VI Return of the Snark: The X-Stern and Sonar

A WHOLE SERIES OF NEW, FAST, POWERFUL SOVIET NUCLEAR TYPES were making their appearance, many armed with antiship cruise missiles. U.S. surface escorts were trying extremely high-powered active sonars as a solution, and they proved very disappointing. Even when the big sonars worked, their "pinging" advertised their presence to a listening sub at ranges far greater than those over which they could pick up an echo. The only hope seemed to be using the nuclear sub itself as an ASW escort, screening and defending the carriers against its Soviet opposite numbers.

This would require extremely high-speed, especially since the escort sub would have to alternate periods of "drifting" and listening at very low speeds with "sprinting" at very high-speeds to keep ahead of the carriers. This was the origin of the *Los Angeles,* SSN-688, laid down in 1972, commissioned in 1976.

The question of speed versus silence produced a major debate and two prototypes, both large and expensive, but still serious compromises in capability. In the end, the fast *Los Angeles* was chosen for series production over the quiet but slow *Glenard P. Lipscomb,* and in spite of its major deficiencies became the standard production class for two decades, with 62 launched between 1974 and 1994, the largest nuclear submarine class in history.

688 was huge—at 6,080 tons displacement, big as some World War II light cruisers. She has *Albacore*'s streamlined bow and stern but is long and cigar-shaped. At 362 feet, she is 110 feet longer than *Skipjack* and twice her displacement; but her 33-foot maximum beam is almost the same. At perhaps 32 knots maximum speed, she buys back every knot given up (and more) by brute force—engine power. Her additional size goes mostly to her GE S6G reactor and twin turbines, giving 30,000 hp; the early Westinghouse S5W, in comparison, gave 15,000.

Sonar and quieting gave *688* the power to detect and pursue Soviet subs, and speed gave her the power to attack them. Maneuverability enabled her to evade their homing torpedoes. For a while, at least, it was considered giving her the superlative maneuverability of Phase I *Albacore*—or better—with the Phase III "X-stern."

The X-Stern

On November 21, 1960, *Albacore* entered Portsmouth for her Phase III conversion and overhaul. During this period she received her X-stern, never duplicated on an American sub, and 10 "dive brakes," a circumferential ring of flaps around her hull, never duplicated on *any* sub. She got Phase I's dorsal rudder back, larger this time, plus a new bow and sonars.

Upon completion in August 1961, she underwent shakedown training in southern waters, a visit to Port Everglades, Florida, then refresher training in New London, Connecticut, and evaluation of her Phase III components by David Taylor Model Basin (DTMB). She remained at Portsmouth from late 1961 until Lt. Cmdr. William P. St. Lawrence relieved Wally Greene as CO. In February 1962, Bill St. Lawrence took her to Fort Lauderdale to conduct stability tests and high-speed-control trials.

And the Snark was back. The new stern restored all of the fantastic maneuverability of Phase I and then some. Along with it came some serious control problems—some anticipated, some surprises. The X-stern consisted of four very, long, thin, "high-aspect-ratio" movable fins with a streamlined "airfoil" cross section. All four fins acted simultaneously in either a turn, a dive, or a combination. This configuration was first used on the Navy's huge new "N"-class blimp of 1951, to provide increased ground clearance at the tail to permit heavy takeoffs at large angles of attack. The X-stern similarly allowed a sub's fins to be longer than the radius of the hull and still not project below her keel, necessary if she was to be able to sit on the bottom (and avoid a headache when dry docking!)

The X-stern increased plane surfaces too, but it was *all* mov-

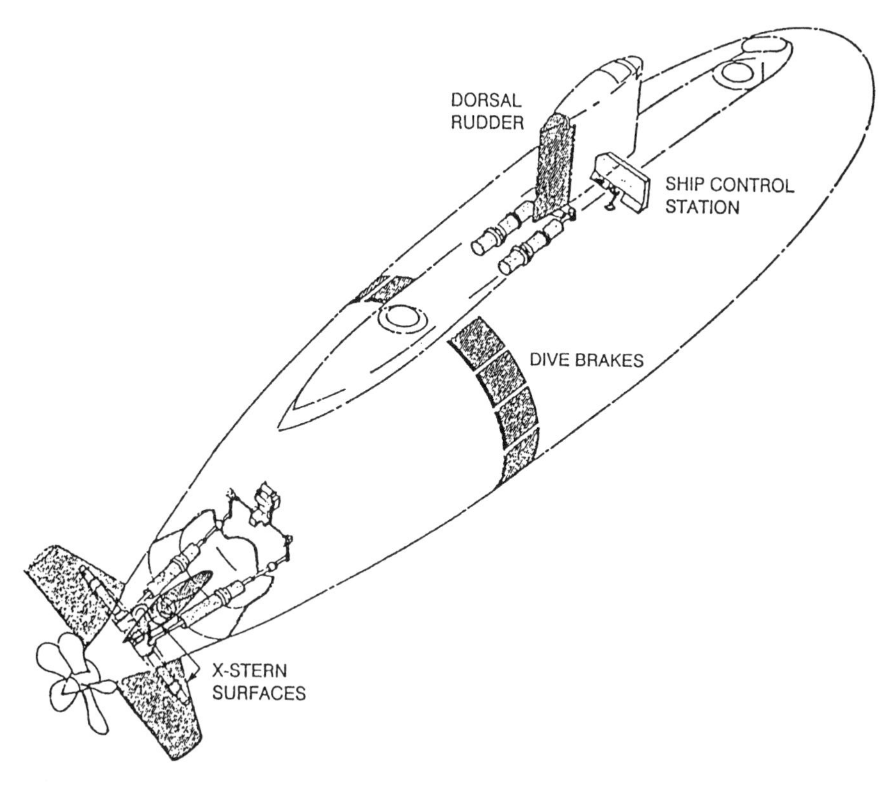

This drawing from Jackson and Allmendinger shows the main features of 569's *Phase III configuration. The X-stern gave her fantastic maneuverability, and consequently, a renewed danger from the snap roll and out-of-control dives. The restored, enlarged dorsal rudder and ten "dive brakes" were intended to counter these. (Courtesy Harry Jackson and Eugene Allmendinger)*

able. All fixed-plane surface and the stability it provided was dispensed with entirely. The X-stern produced tremendous control forces and cut her turning circle nearly in half, from 300 to 165 feet. The dorsal rudder was to counter the vicious snap roll this induced. The dive brakes would counter the renewed potential for throwing herself into an out-of-control dive below her collapse depth. As her engineering officer Ronald Hines put it, "*Albacore*'s pitch rate was twice the amount of her plane rate, so fast that if you got behind, there was no recovery."

However, there was an important safety element in the X-stern. The possibility of a jammed plane causing an out-of-control dive was ever-present. But with the X-stern, all four planes acted as dive planes: Even if two jammed full down, they could be counteracted and the dive

Albacore*'s tail was cut off and replaced three times for Phases II to IV. The idea of giving her variable-configuration tail surfaces was considered but tail replacement was more simple, straightforward and reliable. (Courtesy of Stan Zajechowski)*

angle neutralized by the other two.

With the Phase I stern it had been predicted that as she heeled, rudder angle would become partly a dive plane down angle. With the X-stern this had serious effects. High-speed turns would create a sudden, sharp heel of up to 40 degrees, as Bill St. Lawrence describes it, "left 10 degrees rudder at 25 knots would cause her to snap roll to the left; at that point the X would become a cruciform, and the 10 degrees became a dive plane down angle."

Complex? The system was difficult to visualize and really required a computer, which they had—an analog system. Part of *569*'s job was to prove out its programming; her operators had to learn all the weird effects of the X-stern by experience. "Basically, you had to do it yourself," says St. Lawrence. "We walked her through step by step with an

Phase III Albacore *underway; note the unusual wash created by the X-stern. The X-stern gave* Albacore *tremendous maneuverability but created some control problems. (US Naval Institute)*

infinite number of tests at every speed and rudder angle."

One unpleasant surprise was that there was absolutely no control going astern. In Phase I and II it was bad enough; if her stern was swinging at all when you started backing, she'd turn in the direction of the swing regardless of what you did. When backing submerged was first tried with the X-stern, depth control simply vanished. It was restored only by going ahead flank, to get enough flow over the control surfaces.

"There were many such incidents," Bill St. Lawrence says, "and they were all hairy," as *569* was repeatedly taken deliberately to the limits of control and beyond, to learn her behavior in extreme situations. "We went back one-third at 100 feet and it looked like we had control, so I rang up back two-thirds. The sub started to rotate, I grabbed the periscope with both arms, she went to a 45 degree up angle! And this was under very controlled circumstances, to get data points." Her behavior on the surface was even worse. "It was impossible to dock the ship," St. Lawrence remembers. "She was very tender with those long thin fins. You had absolutely no control backing down with that 11-foot single prop, no rudder because half your control surfaces were out of the water."

He says her enlarged dorsal rudder was effective in countering the snap roll but notes that it was quite large in relation to *569*'s size, and had much more effect on her than on a larger sub.

Ron Hines says that "all the ex-engineers and COs were called down to D.C. for a conference when they were designing the *688* class. The subject was the X-stern. We tried to sell it to them, but they were reluctant because of the problems, in spite of all the safeguards we had built in."

The stability given by the big fixed surfaces of the Phase II cruciform stern was a factor. Says Hines: "They were looking at long, high-speed runs submerged, period. There were concerned about operator fatigue; the planesman leans on his stick and you're past collapse depth before you know what's happening. The recoverability was what they were afraid of."

Norman Friedman writes, "The X-stern was eliminated in the fall of 1969 on the grounds that the associated computer would not be reliable enough." This is in line with Admiral Rickover's long-term distrust of automation and his determination "to keep a man in the loop." Obviously, *Los Angeles* without the X-stern could still snap roll and dive below collapse depth. She would simply have to avoid these things by not turning too tightly or diving too fast.

The Dive Brakes

If the X-stern was successful, these weren't. When the first monoplane dive bombers appeared in the '30s, lacking the drag-inducing struts and wires of their biplane predecessors, they tended to accelerate in a dive—requiring them to pull out early or not at all. The solution was some form of flaps lifted into the slipstream to create drag in a dive. *Albacore* was the only submarine to try this—and with nearly disastrous results. Her ring of 10 flaps were extended out from her hull by hydraulic cylinders and sited in a ring just forward of her aft hatch. Stan Zajechowski calls them "a big expensive waste and mistake," and says that the first time they were used the ship went out of control and to the bottom in 400 feet. "They announced over the speakers 'We hit an air bubble' to calm the men—they wouldn't admit they lost control," he says.

Phase III Albacore *docked at PNS. Note the closed "clamshell" hatches over her bridge, also tremendous span of the X-stern. The X configuration permitted these very long fins; in the conventional cruciform configuration, the lower fin could not exceed the radius of the hull without interfering with dry-docking. (Courtesy of Russell Van Billiard)*

Lt. Cmdr. John McCarthy, later a New Hampshire state representative, says, "We went out to test them and went straight to the bottom. They were not feasible, never used since. First we did a shallow dive—we were off the Isles of Shoals, 250- to 300- foot depth—then a deeper one at high-speed. I was in the forward battery compartment—I remember the collision alarm, the screw backing, the big crunch, our nose in the mud. We hit under the fiberglass sonar dome at a shallow angle, a glancing blow." (Stan remembers seeing the dent when she dry-docked.)

According to Ron Hines, the brakes were poorly conceived. They were too far back on the hull, behind her point of maximum diameter, and thus did not extend beyond her boundary layer. Ineffective for dive control, they continued to be used to create turbulence and thus noise for acoustic tests. This caused trouble too. Ron Hines says that when she was operating off Fort Lauderdale with the Fly Around Body (FAB), a

The "dive brakes" were ten flaps, hydraulically extended away from her hull, to create drag and stop her in an out-of-control dive. Unsuccessful, on one occasion their use threw her out *of control and caused her to hit bottom. (PSM)*

unique remotely controlled maneuverable towed hydrophone, flown like a kite to monitor *569*'s self-noise from various distances and positions, they were ordered to cycle the dive brakes to make noise: "They kicked open, hit full open, made noise, we heard it on the hydrophones. Then we hit the controls to shut 'em—we were losing pressure on the system." Then came the unique and peculiar emergency report over the intercom: "Flooding in the engine room! . . . With hydraulic oil?"

The hydraulic line between the supply tank in the port engine room and one of the dive brake hydraulic cylinders had parted.

Seawater under pressure had flooded back through it, pushing a couple of hundred gallons of oil out of the system's 50-pound pressure relief valve. "The FAB was going good," recalls Hines. "We'd been having trouble with it, so we continued the tests. I don't know if all the brakes shut or not, but that was the last time they were ever operated. We returned and took them off."

In Phase I the stern rudders, thrown full over, served as a dive brake. Ted Davis says this was a surprise; the free-running CalTech model appeared to predict that with full left rudder, the boat would heel, the rudder would become a dive plane, and *569* would head for the bottom. "It didn't," said Davis. "It squatted, lost speed, and then started to spiral down. The doctor [the model designer] was chagrined. The model's reaction time was so short it never had a chance to slow down." But—"just don't leave it on!" Similarly, with the X-stern, throwing the dorsal rudder hard over proved the best way to slow the boat.

However, Bill St. Lawrence says, "The dive brakes were an effective way to get speed off the ship, but it was an expensive sacrifice to get it. They were big, bulky, required structural reinforcement. I'd exercise 'em to make sure they worked, I never had any problems with them."

His own brainchild was the "Jolly Green Giant." While visiting New Hampshire's Pease AFB, he watched a B-47 land, braked by a big green drag parachute. He prevailed upon the Air Force to let him borrow one which was mounted on top of the sail. St. Lawrence theorized that in that position its drag would not only slow the boat down, but pull its nose up and lessen the down angle. Unfortunately, it ripped off on its third or fourth try.

Captain St. Lawrence also curses some of the secondhand equipment *Albacore* made do with, including the "worst in the Navy" radio. She was, of course, required to send in a check report every six hours, or else search and rescue proceedings would be initiated. The problem was in passing through the Cape Cod Canal, where the high sides of the cut prevented radio contact with her base. When they dropped the pilot at the end of the canal, St. Lawrence says "I gave him a dollar and asked him to phone in and say we were all right."

Sonar

The effect of high submerged speed on underwater sound was of interest to DTMB from the beginning. Relatively minor at first, it grew steadily in importance. Some would argue very persuasively that *Albacore*'s absolutely most important contribution—to the United

States and to world peace—was in the realm of sonar.

The first challenge the U.S. Navy posed for the Soviets was finding it, in the wastes of the ocean. To solve this, the Soviets created an extensive ocean surveillance system, using many different vehicles including radar satellites. Had the latter worked, there was the thought that they might be able to strike U.S. carrier forces with ballistic missiles from the Russian homeland. They didn't; but the Soviets searched eternally for some scientific breakthrough that would "make the oceans transparent" and reveal the presence of U.S. submarines as well as surface ships, perhaps by their infrared signatures or satellites carrying blue-green lasers. They never found it; but the United States did. During World War II, scientists from the Woods Hole Oceanographic Laboratories led by Maurice Ewing discovered the existence of sound channels in the ocean, permitting the detection of low-frequency sound thousands of miles from its point of origin. In 1949, the Office of Naval Research opened a sound research facility on Eleuthera Island; there Ewing's associate Lamar Worzel demonstrated to the Navy the tracking of submarines at distances of many miles. In 1950, the CUW's Panel on Low Frequency Sonar made its recommendation that led to the creation of SOSUS—Sound Surveillance System—a network of fixed seabed sonar arrays, 1,000 feet across, mounting 40 hydrophones each, covering the North Atlantic and capable, by the late 1950s, of detecting and tracking every Soviet submarine at sea.

The next advance was to make this technology available to the individual submarine itself, giving it the ability to detect and home in silently on its enemy from great distances. The ability to exploit the power of the new sonars really required the speed and endurance of the nuclear submarine, however. But the GHG was the ancestor of the huge bow-mounted spherical hydroplane arrays around which the *Tullibee* and *Thresher* were designed.

A later step was to give the submarine its own portable SOSUS—towed passive arrays many feet long, of incredible range and sensitivity. The *Los Angeles* was the first sub designed for a towed array and totally passive operation, detecting and tracking her opponents without revealing her presence by "pinging."

The bow spherical array and towed array were first conceived at David Taylor and tested on the *Albacore* under the leadership of Marvin Lasky of the DTMB Acoustics Department. Lasky says that many other people made vital contributions: "It was a team effort in

One area where Albacore *contributed very significantly to the* Los Angeles-*class was in the area of sonar, giving them great superiority to the Soviet sub force in this absolutely vital area.* Albacore *contributed to the development of many aspects of submarine sonar from the beginning of her career. Could the small tripod aft of the sonar dome be for mounting a television camera to observe hydrodynamic effects such as cavitation and turbulence caused by the dome? (PSM)*

which we all marched together," including the Navy men of *569*.

Acoustics research on the *Albacore* was always double-sided: Study of her radiated noise led both to methods of quieting her and to detecting the other submarines—the measuring instruments invented to study her self-noise became ASW sensors. Lasky says, "We worked in partnership with Underwater Sound Labs—every time we had a quiet hydrophone, we gave it to them." Quieting the *Albacore* also improved the performance of her own sonar dramatically.

Before *569* was even to receive a sonar, she was to be given a dummy sonar dome to study water flow and cavitation around it. In fact the JT sonar that earned Ted Davis's curses eventually went inside

1/4-scale mock-up of Albacore's *Phase III bow sonar.* Albacore *was the first to mount her sonar in a "conformal" bow dome (following the curve of the hull itself). In this position it was furthest from the submarine's own sources of screw and machinery noise. (PSM)*

it. A TV camera was rigged to observe water flow around it; a strobe light permitted still photos of the cavitation bubbles forming, causing flow noise. This video system was used on the *Nautilus* on its transit to the North Pole to view the underside of the ice sheet.

Eliminating all other sources of noise, removing the prop, and towing *569* at 10 knots, they discovered the sail was a major source of noise. It created an irregular wake that hit the rear control surfaces and prop. Lasky suggested removing *Albacore*'s sail and replacing the periscope with a towed TV camera. Thomas Gibbons, head of the

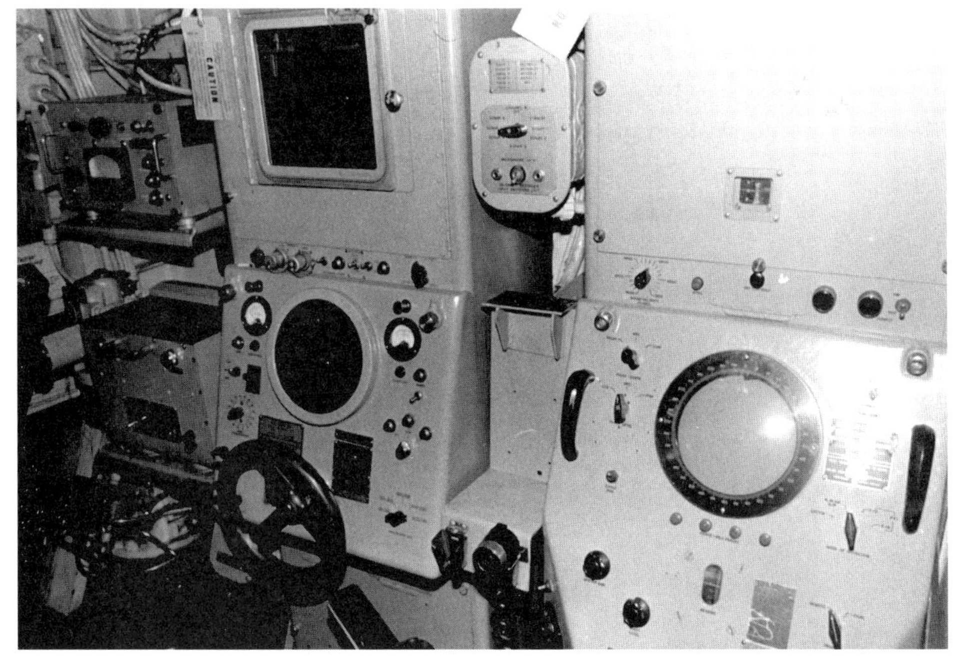

Albacore*'s sonar station, located in her control compartment.* Albacore *went from* no *sonar as built to demonstrating the concepts of bow and towed sonar arrays on which much of our nuclear submarines' superiority in the 1970s and 1980s was based. (Largess)*

Albacore towing experiments, had worked out the necessary technology for this during the war on Project General, a towed paravane to destroy incoming torpedoes. Later, this same system was adapted for towed sonar, and is the basis for it to this day.

The next step was to make the bow itself a water-filled sonar dome. Lasky attributes this idea to Howell Russell, senior projects manager in the DTMB acoustics lab. "Howie Russell was a person of great intuitive mechanical ability," Lasky recalls. "He was invaluable. He gave us an emissary to the shipyard people; he was the person capable of translating the language of us 'Washingtonians' to the Portsmouth workers and turning our desires into practice."

Howell Russell describes the creation of the first bow sonar: "Our object was to create a quiet sonar platform as far away from sources of noise as possible. We cut 6 feet of the steel bow off, took a mold of this, and the shipyard used this to create an identical bow of fiberglass—woven cloth and resin. We installed an early model sonar in the free-flooding area

This photo of the Thresher *under construction at Portsmouth. At the heart of her design, as important as her reactor and streamlined hull is her huge, spherical bow sonar. It reveals the primary role of all US attack submarines to come: killers of enemy submarines. (Naval Institute)*

behind it. All new boats have it, but it was a first for *Albacore*."

During Phase III, *569* received a new, active BQS-4A bow sonar and the American version of GHG, the BQR-2B. In March 1962 these were given a breadboard version of DIMUS (Digital Multi-Beam Steering), invented by Dr. Victor C. Anderson of the Marine Physical Laboratory of San Diego in 1951. DIMUS used an omnidirectional spherical array of 24 hydrophones. Frank Andrews says, "DIMUS was a method of processing sound coming into the array, a narrow beam

Los Angeles *was designed, at least partly, to serve as a high-speed anti-submarine escort to US carrier forces. It was also designed as the first submarine to operate in the "passive" sonar mode, listening silently without generating "active" sound impulses for echo-ranging which would reveal its own presence. This photo shows the long external tube-like housing for its towed sonar array, the key to its incredibly sensitive listening powers. This idea was first tested aboard* Albacore *by its inventor, DTMB scientist Marvin Lasky. (US Naval Institute)*

distinct from all-around noise. It's like 24 searchlights on all at once, simultaneously looking in all directions, unlike a single beam that's mechanically trained."

Bill St. Lawrence says, "It was the first big sonar experiment. It listened for sound over time in unique frequencies that would become a pat-

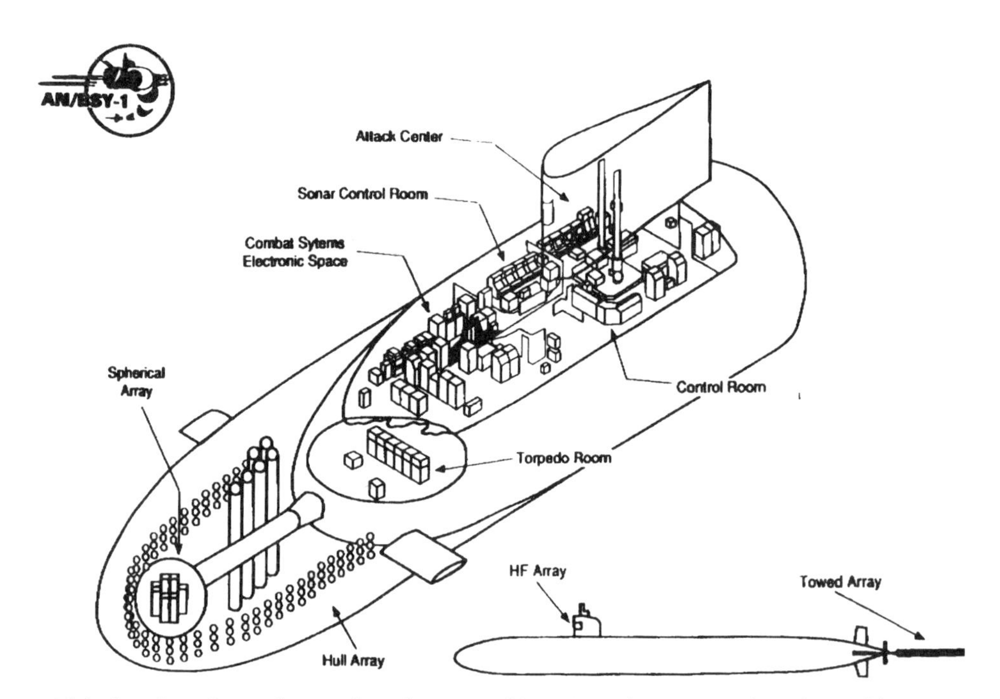

This drawing shows the two key elements of Los Angeles' *sonar, the spherical bow sonar and towed hydrophone array. Both concepts were tried first on* Albacore *and refined over many years of testing aboard her. It was the American discovery of the "low frequency sound channel" permitting the detection of submarines at incredible ranges, which largely solved the Soviet undersea threat for many years. (US Naval Institute)*

tern. Imagine a sub running quiet, but there's guys walking, the cook drops a pan. Over a period of time these little things repeated on the same bearing would produce a coherent track. It was the basis of all our sonars, gave a nuclear sub the passive capability of the SOSUS network."

Between April and November 1962, DIMUS was tested under the direction of the San Diego Marine Physical Lab, in the Boston and Narragansett Bay areas. Along with DIMUS, *Albacore* tested a "sonar receiving array arm" for the Naval Electronics Lab, did vibration analysis for DTMB, and evaluated the effects of bottom contours on torpedo sonars for the Naval Underwater Ordnance Station.

At first DIMUS was met with lack of interest from the Navy and even opposition—as was the towed array. Marvin Lasky, who holds the patent for the latter, says, "It was first used as a method for measuring near-field noise, not as a sonar. It was a simple hydrophone streamed from the sail, first at 15, then 30, then 45 feet—we were careful not to get it tangled in the screw." Howell Russell recalls the first time they tried it, listening on the headphones and hearing gunshots and screams: "Bang! Bang! Aaaaaa!" It was a John Wayne movie playing in the crew's quarters.

688's Debt to Albacore

Los Angeles received *Albacore* speed but not the pure teardrop hull; DIMUS spherical bow sonar and the towed array, pioneered by *Albacore,* but not the X-stern and full *Albacore* maneuverability. The size of *688*'s sail was again reduced, to cut drag, which cost her under-ice and some intelligence-gathering capability. Another element *Los Angeles* lost was diving depth, reportedly only 1,000 feet (or 950 feet, according to Patrick Tyler's *Running Critical*) compared to *Thresher*'s 1,300. This was due to the weight cost of her huge power plant, paid for by reducing pressure hull weight. The Navy hoped to buy back this lost diving depth with HY-130 steel. The Navy's second experimental submarine, the deep-diving AGSS-555 *Dolphin,* also built at Portsmouth, commissioned in 1968, was built of HY-80 but later tested HY-130 components, but the new steel was not ready in time. Here, too, the Soviets pushed ahead, with their titanium-hulled *Alfa* class built from 1970 to 1983, capable of 42 knots and 2,500-foot diving depth.

And here, too, abandonment of the X-stern cost the *Los Angeles*. Speed, maneuverability, quietness, safety, and diving depth are all interrelated. The deeper a sub can dive, and the greater the pressure of the sea upon her, the faster she can go without cavitation, creating bubbles and noise. The X-stern, with its relative freedom from loss of control due to jammed planes, would have given *688* a greater safety cushion for high-speed maneuvers at great depths, all the more necessary due to her weaker pressure hull. But you can't have everything.

Albacore in drydock with the x-stern and about one-third *of her length cut off to permit installation of her twin electric motors. (PSM)*

VII Phase IV: 30 Knots Plus—Submerged

ON DECEMBER 17, 1962, *ALBACORE* ENTERED PORTSMOUTH NAVAL Shipyard to begin her most extensive reconstruction, the Phase IV conversion. For their labors, her crew won her the Navy's Battle Efficiency Pennant, allowing her to display the coveted white "E" for overall readiness to perform her assigned mission. On August 15, 1963, Lt. Cmdr. Roy M. Springer Jr. relieved Bill St. Lawrence. The conversion was completed in March 1965. *Albacore* received:

1. Her contra-rotating twin props and concentric shafting;
2. A second main electric propulsion motor, raising her horsepower on battery to 15,000;
3. Her unusual "silver-zinc" battery;
4. An emergency recovery and main ballast tank blow system;
5. A "vernier mode" plane control system;
6. Her BQS-4 and BQR-2B sonars, a new combined radio antenna system, and other new electronics and control equipment.

During Phase IV, *Albacore* became the prototype for a submarine capable of submerged speeds substantially higher than 30 knots. No U.S. nuclear boat until the SSN-21 *Seawolf* (SSN-21, for "twenty-first century"), authorized for FY 1989, was ever capable of such speeds. Indeed, no other diesel boat ever broke 30 knots.

In addition to her second GE 7,500 hp motor, 60 feet of concentric shafting, and twin props, 20 percent of *Albacore*'s double hull was removed and replaced with a new single-hulled stern. Stan Zajechowski says: "They cut the boat in half after the engines, built a new stern with no tanks—she lost one ballast and one fuel tank. We took the two engines out, and then worked seven days a week, 12 hours a day, to rebuild them completely. Then they sat in a warehouse for 1 1/2 years."

The contra-props had been considered and planned for since Series 58. Besides offering a straightforward way of putting all her new power into the water, they offered many theoretical advantages. The first was propulsive efficiency. The forward screw imparts a rotational force to its wake, wasting some propulsive power. The aft screw recovers this rotational force. Norman Friedman says that when contra-props were considered for the *Thresher* in 1956, they were expected to yield 1.5 knots more, the same as would be gained from removing the sail.

A second, somewhat more complex issue is noise. It has been suggested that, as with the 1959 14-foot prop, the twin props made it possible to put the same amount of power into the water as from the single 11-foot prop, but at lower propeller speeds, thus decreasing cavitation.

Cavitation—the formation of bubbles on the low-pressure forward surfaces of the props—can be reduced by going deep, where the ambient water pressure of the sea is higher. But the blade rate could clearly be heard below cavitation depth. Contra-props could eliminate blade rate by using two props of smaller diameter, turning within the uniform, smooth flow of the hull, and not projecting far enough to hit the sail's wake. *Albacore*'s installation used a 10-foot-diameter, seven-bladed forward prop and an 8-foot, six-bladed aft prop, driven by two concentric shafts. The outer, hollow one was 28 inches in diameter; the inner one, 15 inches. Captain Springer was *Albacore*'s CO while the props were at their first setting, a distance of 10 feet apart. Later, they were moved to 5 feet to discover at what distance they reached their greatest efficiency. TV cameras were rigged to observe the cavitation effect of the props at each setting.

A third noise problem to which contra-props offered a possible solution was gear whine, produced inevitably by the geared turbine drive used in almost all the nuclear boats. One solution tried was the big, bulky turboelectric drive power plants used on the slow but super-quiet *Tullibee* and *Glenard P. Lipscomb*. Another solution was bigger steam turbines, able to turn slowly enough to drive the props directly, but without resorting to reduction gears to slow them down. But direct-drive props would still have to turn faster, eliminating gear noise at the cost of some more prop noise and less efficiency.

As part of her "Phase IV" conversion, Albacore *received a second 7,500 HP electric motor for submerged propulsion. These are the rotating armatures for her two motors being made ready for installation in 1963. (PSM)*

Captain Springer says *Albacore*'s contra-props performed well, with more power transmitted and delayed onset of cavitation.

The silver-zinc battery was not quite unique to *Albacore*. It was temporarily used on the deep-diving *Dolphin* and possibly on the Soviet *Juliet*-class diesel anti-carrier missile subs of the 1960s, probably to give greater submerged endurance, rather than speed. It was used on the TV-1a test vehicle for the *Moray,* the midget, two-man "fighter" submarine repeatedly urged by Admiral Momsen, with aircraftlike controls and extensive automation. Silver-zinc batteries have been used extensively on U.S. and European homing torpedoes. They seem ideal for torpedoes with great power storage and high rates of discharge.

The silver-zinc battery had many times the energy storage capacity of the Exide lead-acid battery it replaced.

The Soviet Alfa-*class submarine, with its incredible performance, 42 knot submerged speed and reported 2,500 foot diving depth, represents the truest fulfillment of the potentialities revealed by the* Albacore *in a nuclear submarine. An incredibly fast, maneuverable, deep-diving boat of very small size with extensive automation permitting a minuscule crew of 27 men, she was intended as an anti-submarine "fighter plane", outperforming her opponents in close combat. Reaching service in 1971, she was phenomenally expensive and complex; only seven were built. (NHC)*

Charging this monster at sea with the little pancake diesels was interminable. As Captain Springer put it, "The battery had such a high capacity, it required long charges with the inefficient and undersized engines. It was a lot quicker to get a charge from the generator plant in Portsmouth or a sub tender. You'd do a few hours of testing, then spend a few days charging the battery."

In March 1965, the Phase IV conversion was complete. The last week of that month was used for refresher training for officers and crew. April was spent in preparation for deployment to Key West, where *Albacore* conducted various stability, control, and shaft-vibration trials from May to July. During this period she again set the world's record for highest submerged speed.

Captain Springer describes a tense moment the first time she went to maximum battery current on the silver-zinc, "faster than any submarine had gone before." Her main circuit breaker tripped "with a tremendous bang." This cut out current to the AC motor generators, leaving her without internal power and in total darkness. With her hydraulic pumps out, she went into emergency mode, with high-pressure fluid being supplied from the system accumulators, so her controls remained operable. However, for about eight seconds, until her dry cell emergency lighting came on, her instruments were totally invisible to the officer at the "stick"—plenty of time for *569* to go below her collapse depth out of control.

The diving officer, however, suddenly noticed the hissing of the hydraulic oil flowing through the circuits of his control stand as he moved the stick. He had never been consciously aware of hearing this sound before. But he immediately realized that he could tell what he was doing by the changes in sound as he adjusted the stern planes up and down. These subconsciously familiar sounds enabled him to keep *Albacore* on a stable course for the necessary few seconds before the emergency lights came back on.

Captain Springer says: "We trained for that kind of experience. It was tense, but well handled. Safety training was our primary concern as a crew."

On August 1, 1965, *569* returned to Portsmouth Naval Shipyard to be outfitted for photographic and vibration trials, using TV cameras mounted on tripods outside the hull. On August 25, she arrived at Port Everglades, Florida, and began the photographic studies of cavitation by the contra-props. During September, self-noise and towflex trials were conducted. She returned to Portsmouth on October 8, 1965, concluding the first stage of Phase IV, which contributed data regarding the operational feasibility of the silver-zinc battery (then planned for the *Barbel* but ultimately not finally included), as well as data on submarine control and maneuverability at her again unprecedented high-speeds.

Another novel feature tested was a "vernier" system of rudder and dive plane control. According to David Merriman, "Whenever *Albacore* exceeded a set speed, the control surface servos were automatically fed reduced signals from the control wheels. This allowed smaller control surface movements with normal throws of the helm and planes control wheels, thus preventing overcontrol at high-speed."

Albacore also tested various components of a new system for blowing water ballast. The first component was a push-button ballast control panel, a simple switch and indicator board controlling solenoid-operated

Albacore *in dry-dock with USS* Dolphin, *AGSS-555, on her left in 1970.* Dolphin, *completed by PNS in 1967, was the second purely experimental submarine. Where* Albacore *was designed to investigate control at high speed,* Dolphin *was designed to explore metals and materials capable of withstanding the extreme pressures at great depths. (PSM)*

remote valves located at the tanks themselves. This replaced the old system of leading the air manifolds through the control room, covered with a jumble of mechanical valves and gauges. This new panel was developed for the *Barbel,* saving scarce control room space in this *Albacore*-sized boat that had to incorporate three diesels and combat equipment.

However, the systems tested in 1965 were developed in response to the loss of the *Thresher* with 129 men on April 10, 1963. This tragedy was a severe shock to the Navy. *Thresher* was the result of years of refinement of nuclear submarine design, and the first true general-purpose submarine warship, duplicated in a production class of 14 boats. On her deep diving trials off New England, *Thresher* had suddenly lost nuclear reactor power. Normally, a modern submarine does not attempt

The loss of the Thresher *in 1963 led to the SUBSAFE program. Here,* Albacore *tests a new 3,000 psi emergency ballast-blow system, which would permit a submarine to surface safely even if all power was lost. (Courtesy of Norman Bower)*

to surface by blowing her tanks against the tremendous pressure of the sea at great depths. Instead, she rises on her planes, using their lift to bring herself to periscope depth, and blows her tanks there where the pressure is much less. At the time it was speculated that *Thresher* could not "plane up" due to her loss of power and her ballast blow system lacked sufficient pressure to force the sea out of her tanks at her unprecedented test depth of more than 1,000 feet.

In fact, she should have been able to blow her tanks. It is conjectured that the sequence of events causing her loss began with a pipe carrying seawater, e.g. a condenser pipe; many instances of defective silver-brazing in such pipes due to an initial design deficiency had already occurred. Flooding could have shorted out the reactor controls, which shut down automatically as a fail-safe maneuver, leaving the crew with no means of immediately restarting the reactor, and the *Thresher* with no motive power, unable to plane up.

What prevented her from blowing tanks was probably this: high pressure air was led through long narrow pipes to the tanks. They inevitably contained some moisture. As a gas expands it cools adiabatically. When this happened in the pipes, the moisture froze into ice and clogged the pipes. Friedman says, "It later turned out that all components of the system had been tested individually but not the system as a whole .... The basic design was not at fault, but it had to be modified to preclude any repetition. The new program was called SUBSAFE. It delayed attack submarine deliveries by several years."

Some parts of SUBSAFE involved reactor control, but those involving emergency main ballast tank blow at very high pressures were developed and tested by the *Albacore*. David Merriman says, ".... electro-pneumatic valves blew automatically if electrical power was lost, insuring that if either the 'blow' switch was thrown or if there was a major loss of ship's power, there would be a quick blowing of ballast and the boat would quickly come to the surface." However, Russell Van Billiard says this is not true; it was not completely automatic. "You still had to do it - there were two levers in the control room you had to activate (panic switch situation!)." In "emergency" mode, *Albacore*'s instantly slammed 3,000 psi air pressure into her tanks.

According to Ron Hines, when testing the system, "stationary blows caused tremendous lists. She'd be captured in her own air bubble, unstable, and any outside force would cause her to roll. It was best to have 10 knots or more way on. Her superstructure wouldn't drain right. We had to cut big limber holes so it'd drain and not overbalance the ship. And when she listed coming up, the flat surface of her sail would catch and pull her even further over."

The "Fixed Crew": Service Aboard the Albacore

Captain Springer emphasized training for safety. The reason was

Albacore's many modifications. As Ron Hines said: "*Albacore* in the shipyard, in the dry dock, was probably the most unpredictable submarine ever built. You could not ever tell when you flooded that dry dock whether she was going to lean over 10 degrees or what she was going to do. When we went to sea, on the first dive, we didn't know either."

At this point the name "Snark" had been forgotten, if it was not just a private joke between Ken Davidson and Ken Gummerson, but *Albacore*'s behavior was as elusive and Snarkish as ever. She needed a resourceful, skilled, and experienced crew who knew her quirks in and out. Ron Hines says, "The crew needed major expertise, making sure it would run, making sure all the projects tried were installed correctly and did not interfere with the regular ship's operating systems. Every time we went to sea we had something experimental. We'd operate through a project, then at Portsmouth we'd tear the ship apart and get it ready for the next project."

This required a "fixed crew" essentially permanently assigned to the ship. Hines says the Navy "put a hold" on the *Albacore*'s crew back in the early 1960s. Instead of the normal 3 years maximum assignment, people would stay much longer—Norman Bowers 13 years, Stan Zajechowski 16 years. (Stan says he had the most time on *569* of anybody. He stayed, no request for transfer, thinking his last request would be good shore duty in Europe. Instead, he was sent to San Diego and then to Vietnam, where he spent a year on an oceangoing tug off Da Nang!) Stan says he misses "the camaraderie, the brotherhood . . . civilians just behave differently."

There were advantages; as Bill St. Lawrence puts it, "*Albacore* was good duty. We were home-ported in Portsmouth, spent most winters in Fort Lauderdale on trials. It meant a stable home life, something you often miss in the Navy."

Says Roy Springer, "One nice thing—the crew were aboard because they wanted to be, they had fathers at the shipyard, where we were home-ported. It was an extremely close crew; I've never been associated with one like it. We operated at Key West one summer for seven weeks. We had *no one* involved in Shore Patrol reports—the squadron commander was astounded!"

Yet even before the fixed crew, *Albacore* was showing certain characteristics that stuck with her throughout her life: the tremendous rapport between Navy men and scientists, the competence and closeness of the crew, the commitment to their mission, the coolness and humor with which they dealt with *Albacore*'s Snarkish surprises.

According to Norman Bower, these shots show Albacore *in extreme distress, caught in a hurricane in the late 1960s with an exhausted battery. The combination of heavy seas and* Albacore's *very poor surface characteristics left her taking water over her bridge and in through her main engine air induction at the top of her sail. Reportedly, she came very close to being lost. (Courtesy of Norman Bower)*

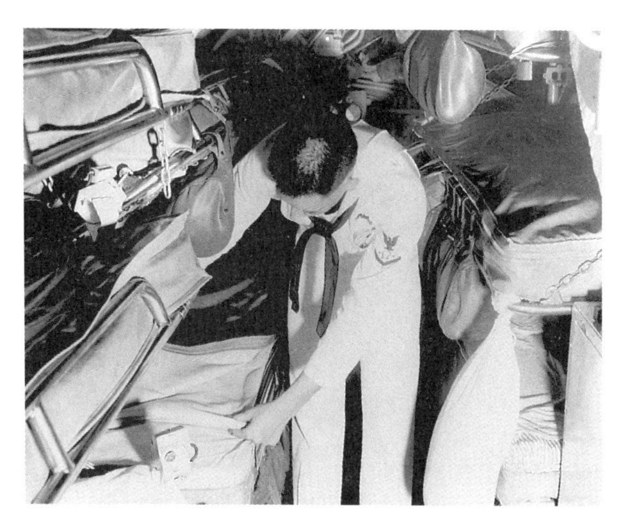

These often-published cruise shots from early in her career show a fresh, youthful crew not much like the experienced old-timers who make up Albacore's *"fixed crew" during the 1960s. Because her behavior after her many modifications was totally unpredictable, it was necessary to have a seasoned crew who knew her thoroughly. (National Archives)*

Lando Zech recalls: "A fine ship—it was clear to all this was the ship of the future. All knew that nuclear subs would incorporate her lessons.

"Danger? The greatest strength you had was the crew—they could handle anything. Not overconfident; we were operating on the threshold. And she was really a pretty stable ship—you'd hear that emergency backing bell, slow her, hands off, and you knew she'd right herself.

"The people loved that ship—the enginemen babied those pancake diesels. The finest electricians, even the cooks put up with those angles! People wanted to stay aboard. You really did feel kinda special—you felt a great bond with your sailors because you all knew you were doing something important.

"We had a fire in the battery ventilation system—the sailors responded immediately, donned gas masks, dove into the battery compartment. They jumped right to the fore; that's where experience really tells.

"A tremendous bond of brotherhood—a special bond on the *Albacore*—you were supposed to do things you'd never done before."

He does not say so, but we can guess that an important part of the equation was Lando Zech. Certainly each of the skippers brought something unique to her; many have stated that service aboard *Albacore* was the high point of their careers. Her COs were plainly ardent submariners, fascinated by the potential of the *Albacore,* stimulated by the challenge and independence of pure research into the nature of the submarine and her environment.

"It was the closest thing to being John Paul Jones," Bill St. Lawrence says. "They gave me a ship, gave me a crew, and told me to get the job done. There was no one looking over your shoulder. They left me free to act, in the best interests of the crew, of the mission. It was a pretty heady experience for a 33-year-old man."

About Ken Gummerson, Frank Andrews said, "He had a genius for making things work"; about Jon Boyes, "He had a flair for imagining things, doing new things, getting people going in new directions."

Jon Boyes says of Ken Gummerson, "He was one of the great naval people of the day, very technically proficient, a bridge between the engineers who designed it and the operators who were to fly it. Gummerson did an enormous job completing her, not getting much glory. He made me see what was possible. I was blessed with a great wardroom and crew—there was sort of a synergism.

"People advised me: First, *Albacore* is the worst assignment you can get! Second, if you really do achieve anything unique with her, well, that makes you a maverick; it'll spoil your Navy career! Third, they'll say afterwards that whatever it was you achieved with the *Albacore,* it wasn't what she could and should have achieved. *You* spoiled the *Albacore*! But the advice I took—the advice from the best and brightest—was hang in there and do it."

Radioman Ron Poloske remembers *569* as great duty for a married man; out of Portsmouth at 7 A.M., in at 4. Most of the crew were married, but it was pretty boring for a 20-year-old who had "joined the Navy to see the world. But the Chief of the Boat was trying to get his daughters married off—three of 'em—so he'd have us single men over all the time."

They lived in the barracks, ate lunch and dinner with the shipyard workers, and had good rapport with them. "The *Nautilus* crew were stuck-up SOBs," said Poloske, "wouldn't let the yardbirds aboard, wouldn't eat with 'em."

It was exciting to go aboard, according to Poloske: "With *Albacore* on the surface, you'd have your head in the commode—seasick!—until we pulled the plug. She'd start rocking at the dock when a lobster boat went by, because of that round hull." When they went to Bermuda, Bill Rae, "a quiet man," told them to bring civilian clothes as they could stay in the hotels. He remembers meeting the old man and the officers and having a drink with them there.

Howell Russell and New Hampshire State Rep. John McCarthy remember *Albacore*'s sixth commanding officer, Lt. Cmdr. Wallace R. Greene, with particular affection. During his command, *569* seems to have acquired her only service nickname: "The crew used to call her 'Wally and His Little Red Wagon.'" says McCarthy. "He was an ex-enlisted man, either a 1st class signalman or 1st class quartermaster. He rose from the ranks, made full commander before he retired—he was division commander of Submarine Squadron 6 in Norfolk when the *Scorpion* went down.

"He was a big man. When we were going to Bermuda, Wally would say to Bob McConnell—one of the two other men aboard over six feet—'Bring an extra set of whites.' He had a habit of going ashore the first evening dressed as an enlisted man. He'd go to one of the clubs, smoking a big cigar. And he'd say, 'The first SOB who calls me Captain or says Sir, I'll deck him!' He just wanted to be his old self and play sailor for one night.

This rapport and camaraderie included the civilian scientists, many of whom dedicated large parts of their professional careers to *Albacore,* went to sea aboard her over a period of many years, and were seen by the crew as shipmates. Marvin Lasky says:

"Our Model Basin civilian crew had the best cooperation not only from the officers but also from the enlisted men, especially in the forward room, where we installed the bulk of our gear. They not only 'hot-bunked' elsewhere but also cheerfully gave up their bunks to the Model Basin research personnel and helped us install our instruments in the superstructure and around the propellers. The Chief of the Boat on numerous occasions kept us from harm when he himself helped install outboard hydrophones on struts near the razor-sharp tips of the propeller. Crewmen also put on scuba gear and dived to remove burr damage to the tips of the screws caused by hitting stray logs and detritus coming in and out of port.

"Howell Russell's magic in materializing help from PNS was also proverbial. The hero list of the *Albacore* includes Cmdr. Richard Dzikowski, who wrote the details of the X-stern installation. *Albacore* was a miracle brought to fulfillment by the goodwill, superb team effort of many dedicated individuals. We were blessed by the good fortune to be friends and close companions, that we were not killed by the same misfortune that overtook our friends and sometime shipmates on the *Thresher*."

Many scientists speak of the consideration and respect shown by the COs. Howell Russell remembers, on his departure at the successful conclusion of the Fly-Around-Body (FAB) trials, Captain Springer calling on him with the crew assembled at dockside and at attention: "Howie, say a few words to us."

Dick Stensen of DTMB's Full Scale Trials Branch remembers working on the propeller spacing trials in the 1960s: "A typical day, you'd go aboard at 3 A.M. in wintertime to get under way and out of Portsmouth at slack tide—I remember one of us slipping on the icy deck and going overboard! It was crazy hours, long hours getting data. Long transits on the surface to the test area, dive, run tests. Our people would commute from Washington on a daily basis.

"You knew the crew on a first-name basis; we worked with the ship on a long-term basis like they did. Living accommodations were cramped, mattresses on the wardroom table, under the table, under bunks. You want to go to bed, they're watching the evening movie—you wait, then they're serving the morning meal. Some slept on the deck in the engineering spaces.

Here is Albacore's *official "Praenuntius Futuri" patch, dignified and serious, showing the streamlined albacore fish. And here is her crew's informal "Roadrunner" patch, with its message: "Beep-Beep We're the fastest." Versions of this, as patches, plaques, and wallet cards, go back to* Albacore's *commissioning crew, first with a pre-Warner Bros. realistic roadrunner bird and "Speed Merchants" motto. Plaques were awarded by the skipper on the departure of a particularly valued crewman. At least one civilian scientist received one. (Authors)*

"Warner Bros. gave them permission to use The Roadrunner on a plaque. I was the only civilian to receive it, presented by Captain Kratch at the decommissioning. He was very professional, all business in port. He held competitions between the civilians and the ship's force, basketball games, foot races. Once he brought in a ringer, an officer on reserve duty who'd just run in the Boston Marathon."

One of the major points in Gary Weir's book is the fruitfulness of this all-to-brief period of Navy/scientific confidence and trust. He quotes Marvin Lasky:

"Let me tell you why the civilians were able to control *Albacore*. Because the spillover of scientific effort from World War Two and the respect of the naval officers in charge for civilian expertise in solving naval problems. This has since evaporated."

The Final Speed Record

In November 1965, Roy Springer was relieved by Lt. Cmdr. J.W. Organ. Operations during the following nine months were largely restricted to accumulating acoustical and machinery performance. Intercept Hydrophone Evaluations were conducted in December 1965 and January 1966. *Albacore* entered dry dock on January 13 to 14, 1966, to have her hull cleaned for further standardization trials on February 1 through 4. At this time *Albacore* was again officially reported to have set a new world's submerged speed record. *Jane's Fighting Ships 1967–1968* gives her speed then as 33 knots, probably based on an official figure provided them. However, later official U.S. Navy correspondence refers to her as a "35 knot submarine."

In April 1966 she conducted Radiated Noise and Shaft Vibration trials for DTMB. May through June was mainly filled with acoustic evaluation of towed arrays, though *569* took time off to represent the Navy at the annual Broiler Festival in Belfast, Maine, where, unusually, she was opened to visitors. Training continued in July with an emphasis on photoreconnaissance and simulated torpedo runs.

On August 1, 1966, she entered Portsmouth Naval Shipyard to have the silver-zinc battery replaced. On August 15, Cmdr. Roger H. Kattman relieved Commander Organ, and for the next two months *Albacore* conducted standardization and machinery trials in the Gulf of Maine with the relocated contra-props. On October 10, 1967, she went to Port Everglades and evaluated the Fly Around Body. There she was again opened to visitors and took Fort Lauderdale Navy League members and city officials on a one-day cruise, as she had done previously under Wally Greene. In November 1967 she again went to Atlantic Undersea Test and Evaluation Center (AUTEC) for radiated noise trials.

On January 1, 1968, she entered Portsmouth Naval Shipyard to again relocate the contra-props for her final Phase IV trials, beginning on April 19. After one month she headed south to Tongue-of-the-Ocean, to work with AUTEC and Mobile Noise Barge (MONOB I). Tongue-of-the-Ocean is a body of deep water east of Andros Island in the Bahamas. It is a gigantic undersea canyon, 100 miles by 15 miles and up to 6,000 feet deep, with immensely steep drop-offs forming its side walls and only one access channel.

MONOB I (Mobile Noise Barge), originally built as a self-propelled water barge in 1943, was equipped with transducers that could be lowered to various depths. Civilian-manned, it carried out acoustic trials with Navy ships for many years. In 1968, it was moored off Eleuthera Island, east of Nassau, where *Albacore* also evaluated the "Phase I" FAB.

The Fly-Around Body, as mentioned earlier, was an unusual hydrofoil-shaped tethered body, "flown" by remote control from within the *Albacore*, using hydrodynamic lift to "rise" against the pull of its

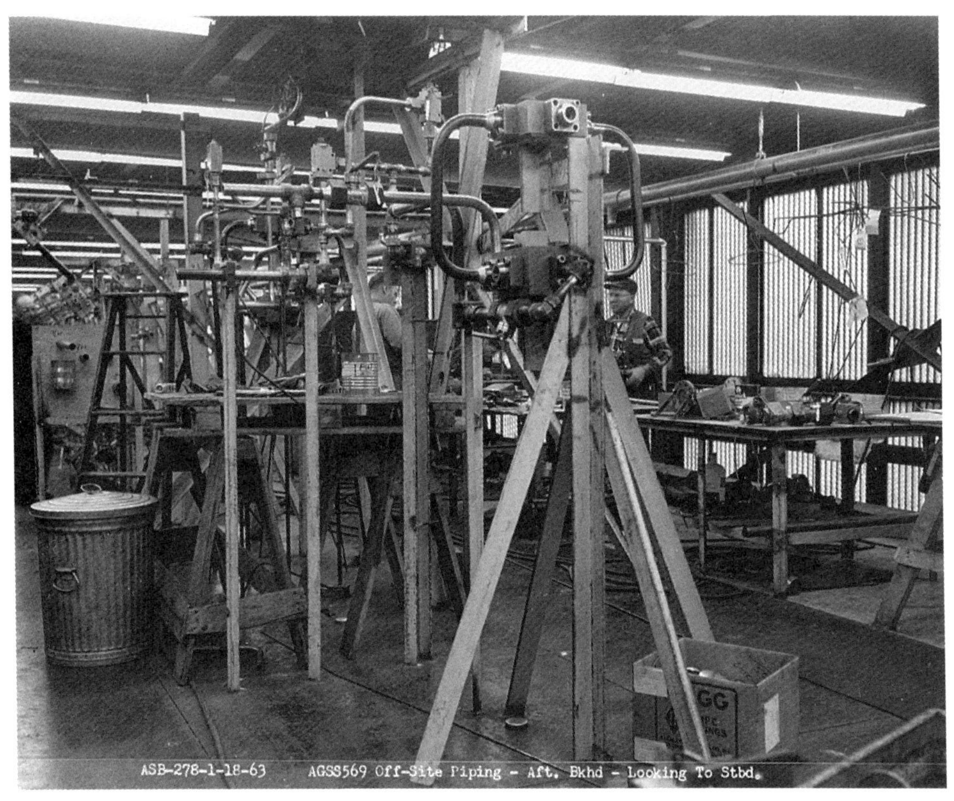

Off-site machinery compartment piping being pre-fabricated for later installation in Phase IV Albacore. *The more pre-fabrication done off-site where plenty of room is available, the better. (PSM)*

tether cable, essentially just like a kite. Howell Russell has described how it was used to monitor *Albacore*'s own "near-field" noise anywhere from 50 to 100 feet away; it could be "flown" out to whatever exact distance one wished to listen. Every nuclear submarine had to undergo periodic sound measurement to eliminate noise problems that might be caused by design or mechanical problems and might be ruining its quietness and thus "stealth." With AUTEC, every submarine had to go to one very expensive instrumented sound range in one specific place for measurement. With the FAB, Lasky hoped to develop a simple, inexpensive piece of technology that any submarine could use to check its noisiness without any serious interruption of its operations.

Albacore returned to Portsmouth Naval Shipyard on August 24, 1968, for removal of AUTEC instrumentation and installation of FAB

The 1/4-scale mock-up of Albacore's *Phase IV machinery compartment. The mock-up is a design tool to assure that everything will fit when built and actually installed in the boat. (PSM)*

Phase II equipment, evaluated in the Gulf of Maine. Back at Portsmouth, *Albacore* went into reduced operating status, pending the results of studies on the feasibility of using her for future research. On July 11, 1969, Commander Kattman was relieved by Cmdr. Thomas E. Poole. During this period, she visited Edgartown on Martha's Vineyard and Rockland, Maine, but remained inactive for the most part until February 2, 1970, when she entered dry dock for overhaul and modifications to prepare her for Project SURPASS.

Phase V: Project SURPASS

Project SURPASS was planned by DTMB (or to give its current name, the Naval Ship Research and Development Center in Carderock,

This peculiar bow installation is for handling the "Fly-Around-Body", a winged kite-like shape towed from the bow which could be steered by remote control. Carrying sonar hydrophones, it was used to listen to the submarine's "self noise" at various distances. (PSM)

Maryland). Phase V grew out of the long-term interest in finding a breakthrough in reducing skin friction through boundary layer control—the breakthrough that Admiral Momsen had confidently predicted would yield 50-knot speeds. This would be achieved by the use of viscous polymer liquids, stored in tanks and expelled from jets to coat the hull at high-speeds.

The 40,000 gallons of fluid, consisting of 0.2 to 0.8 percent of the polymer itself, mixed with fresh water, was stored in three soft tanks. One was in the bow dome, one in the former NFO-2 tank, and a tank was constructed around the space formerly occupied by the dive brakes. The fluid was emitted through two ejector manifolds or rings of jets, one around the bow dome and one replacing the dive brakes aft of the sail. The driving force to eject the fluid was provided by a water ram inlet on the leading edge of the sail. As the ship went ahead, salt water under pressure caused by her forward motion was led from the ram inlet to the bottom of the tanks, forcing the polymer liquid out through the ejector ring jets.

The effect is real and substantial. British naval architect D.K. Brown writes, "Even at concentrations of the order of 10 parts per million . . . reductions of the order of 30 percent in frictional resistance have been demonstrated." A natural substance, guar gum, was tried by the U.S. Navy in 1963. Britain used polyethylene oxide in a trial on HMS *Highburton* in 1968 and achieved a reduction of 17 percent in the horsepower required to achieve the intended speed. The problem is the very large quantities of fluid expended in even brief periods of use. The installation consumes much space, the fluid is relatively expensive (certainly much more than the amount of fuel saved!). It would have to be hoarded for moments of tactical necessity, when a burst of top speed was required. Could it be of real, practical value to a submarine? *Albacore*'s job was to find out.

On April 16, 1971, she left Drydock One at Portsmouth Naval Shipyard, with her modifications complete. On July 22, she commenced sea trials; she went to New London from August 6 to 13 for crew refresher training at the Sub School. On September 25, Cmdr. David A. Kratch relieved Commander Poole. In October, she operated off Provincetown to calibrate her radar and sonar, and on November 8 she commenced at-sea operations in support of Project SURPASS.

Commander Kratch wrote in February 1972 that as of December 31, 1971, preliminary tests showed that a polymer ejection rate of 1,550 gallons per minute resulted in a 9 percent speed increase. Conversely, if the ship remained at the resulting 21 knots, she could maintain it with only 77 percent of the normal horsepower. At this rate, her 40,000 gallons of fluid would last about 26 minutes. It seems apparent that the value of this effect for a nuclear boat would be in permitting brief bursts of higher top speeds for evasion or attack. Did the *Albacore* use it to once again break her own speed record? The answer is no.

The problem was with the GM 16-338 pancake diesels. Commander Kratch wrote, "Their unreliable operation is well documented in the annals of naval materiel history. . . . The number of overhauls have precipitated the following local operating restrictions: (1) Both engines in commission prior to going to sea, and (2) if one engine suffers a casualty at sea, immediately return to port." Each engine casualty in 1971 caused a revision of the trials agenda; enough casualties ultimately brought about the cancellation of Project SURPASS.

And that was it for the *Albacore*. These were the last pancake diesels in existence; there were no more. Phase V testing ended in June 1972. Norman Friedman says there were thoughts of replacing the pancake diesels for a possible Phase VI series of unspecified tests. But this would have required the insertion of a 12-foot midbody section: "This would have added considerable drag and made Phase VI too expensive."

Albacore paid a visit to Bar Harbor in July; then preparations began to deactivate her. On September 1, 1972, she was decommissioned; Rear Adm. Paul Early and Robert A. Frosch, Assistant Secretary of the Navy for Research and Development, spoke at the ceremony. She then joined the mothball fleet at Philadelphia.

Albacore *went to the mothball fleet at the Philadelphia Naval Base upon decommissioning in 1972. Here she is in August 1976, tied up next to* Gearing*-class destroyer* Robert L. Wilson, *and another one-off from the 1950s, the prototype fleet command ship USS* Northampton *(CC-1). (NHC)*

VIII *Saving the* Albacore

EVEN WHEN *ALBACORE* WAS ACCOMPLISHING HER GREATEST achievements, not everyone was enamored with her. ComSubLant was often rumored to be her biggest non-fan; she took funds and crews away from the operational submarine force. For example, ComSubLant E.W. Grenfell discussed the decommissioning of *Albacore* in a memo to the CNO back when she was undergoing her Phase IV conversion in 1963. He noted that she had the last two "notoriously unreliable" pancake diesels; already this "well had run dry." Because her silver-zinc battery required 22 hours on both engine generators for a recharge, *569* was required "to operate with an escort or within sight of land."

Though Phase IV *Albacore* was "designed to make 36 knots," Grenfell notes the perennial dangers of loss of control, and a proposal to fit a second set of small control surfaces for use at high-speed (apparently never carried out). He notes that *Jack* would be available in August 1964 to complete studies of the contra-props.

And, he noted, "The cost of operating *Albacore* is approximately three times higher than that of another diesel submarine"; also, "Six officers and sixty men are engaged in operating a submarine with no weapons delivery capability." In conclusion, he said, "In view of *Albacore*'s questionable future research potential and in light of the safety, materiel, and personnel factors cited above, it is recommended that *Albacore* be decommissioned upon completion of those Phase IV trials deemed absolutely essential for future submarine development."

Still, even in 1972, the hope for more life for *569* was not absolutely dead. Before decommissioning, her last engineer, Lt. Cmdr. Robert L. Ball, compiled a detailed summary of her material condition, addressed to "Prospective Engineer, USS *Albacore,*" advising of this elusive (or should we say Snarkish?) personality of her problems.

He spoke of the pancake diesels: "If you are reading this document, a decision has undoubtedly been arrived at regarding what your replacement engines are to be. It is felt necessary here to state that nearly all of *Albacore*'s material headaches have stemmed directly from the two GM-16/338 diesels she has as of this writing. . . ." Now *569* had "the remaining two in existence in the world, no tooling facility and most of any spare parts left anywhere on the globe in salvage yards. Upon decommissioning there are two engines and a cage full of retrieved parts. These parts would give approximately one year of life to the engines at the present rate of consumption."

He praised her stern shaft seals. These are often mentioned as a reason for the failure of the contra-props on the *Jack,* but this seemed incorrect: "Finally do not become discouraged if on initial sea trials you ship water through the seals. It is a little unnerving but they are good seals and they will work well for you once you start operating."

The battery was gone, removed for use in *Cutlass*. The main motor installation was set up for the silver-zinc battery and must *never* be put in MAX with a lead-acid; the ampere drain would be disastrously high. And the traditional bell of DEAD SLOW provided a minimum of 60 RPM or 8 knots! "This is true surfaced or submerged and means shiphandling quite often includes 'pulsing' (ALL STOP/DEAD SLOW)," he wrote.

He warned about plane jams: "Double plane casualty is nearly zero; single plane casualty a possibility." How to handle it? Always back down, do not overreact, know the indications for plane jam. "If you had unlimited depth capabilities," he wrote, "the ship would pull itself out and end up in a flat spin turn, however to prevent excessive depth excursions at high-speed, operator reaction is necessary." Ideally, full left rudder, back two-thirds and 10 degrees rise would pull out of plane jams. "Remember, if necessary, you still have the dorsal rudder and 3000 psi blow system," he wrote.

He recorded her great achievements: the teardrop hull, HY-80, SUBSAFE, the Emergency Main Ballast Tank Blow System. He noted that she still had the polymer tanks and ejection rings installed.

Now, to bring *Albacore* to operational status again. You may have to battle the Shipyard for updated drawings and plans, he wrote. "This seems a ridiculous matter to have to discuss but experience being the best teacher—necessary.

"Remember, your end product depends on you—fight for everything soundly required for safe operation of your plant using your commanding officer and your funding project office to assist you in gaining approval when needed."

Thus, Commander Ball attempted to pass on the accumulated experiences of two decades of tending and nursing the Snark. His words already had the hollow sound of whistling in the dark—there was to be no engineer, no reactivation. No one read them until *Albacore*'s historian, some 20 years later.

The Next Two Decades

Albacore went to the mothball fleet, the Inactive Ship Reserve Facility at Philadelphia, to join her fellows, the ships that were new and beautiful and advanced in the 1950s. She sat next to the Guppy *Clamagore* for a while; the Guppy seemed to dwarf her, so much more of *Albacore* was under the surface even when docked. She was tied up to the *Northampton,* another custom built from *569*'s era, the prototype fleet command ship, built on the hull of an unfinished World War II *Baltimore*-class heavy cruiser, almost unarmed except for her extensive electronic equipment.

During these years the *Los Angeles* class came to fruition, along with Admiral Gorshkov's new navy, the first real blue-water fleet in Russian history. *688* still kept the nuclear submarine edge for the U.S. Navy during the 1970s, with her speed, quietness, and advanced sonar, including the towed array. But once the Soviets learned about SOSUS and "got religion" about quietness, things began to change. In the 1987/1988 edition of *Jane's*, the editor remarked:

"Since the commissioning of the first of this [*688*] class in 1976 the Soviet Navy has introduced four classes of nuclear attack submarines, at least three with an increased diving depth, all with the far higher power density than *Los Angeles,* and with a much superior armament . . . all the result of innovation and imagination. . . . Because of conformism, conservatism, and complacency the U.S. Navy will not have a radically new design of submarine at sea until 1994, USS *Seawolf*."

In other words, with their higher speeds, deeper diving depths, great structural strength, heavier armament, and good if not better sonar and quieting, Soviet subs might just have gained the edge, especially in "melee warfare." In 1987, Lt. David Nylen wrote: "Traditionally the role of SSN was similar to that of a skilled hunter with a high-powered rifle. The hunter found his prey and dispatched the hapless victim with one shot fired from a long distance, which kept the shooter relatively safe from attack. In the future, the US SSN designed for such long-range hunting will find itself in a melee at close range, much like a knife fight in a dark alley."

These were the lessons Jon Boyes and the *Albacore* tried to teach in the 1950s; suddenly it seemed we had forgotten them at our peril. Besides, our boats could not be simply equal in quality; we depended on their technological superiority to counter Soviet superiority in numbers. In 1985 we had 100 nuclear and 6 diesel attack boats, versus 130 nuclear and 250 diesel Soviet attack and anti-ship-missile boats. And note that while during these years that the Dutch, Japanese, and most of all the Soviets, in their numerous *Kilo* class sold around the world, directly copied the *Barbel* and *Albacore* in the form of small, inexpensive, fast, teardrop-hulled diesel boats, the United States let its diesel boats dwindle away to nothing—though not without infinite debate, of course.

Meanwhile, Portsmouth Naval Shipyard engineer Russell Van Billiard was making frequent trips to the Philadelphia Navy Yard in the late 1970s. He noticed *569* tied up there "wasting away—looking forlorn," as he put it. Back in Portsmouth, he began to talk to people about the idea of building some sort of effort to bring her home, as a museum and ship memorial.

But the Navy had other ideas. The Sub-Board of Inspection and Survey of the Naval Inactive Ship Maintenance Facility conducted a survey of *Albacore* December 10–11, 1979. They found that lacking nuclear power, a snorkel, any weapons capability, and more than one periscope, she was unfit for further service. She would require an additional section in the forward compartment to accommodate torpedoes and tubes, and an additional section in the control room to take an attack periscope and fire control equipment. They noted that her acquisition cost was $5.3 million, her replacement cost $55 million, and her scrap value $178,000. Her activation work package would cost $6.6 million, and as for full-scale modernization—who could guess? They observed that when the *Tang*-class *Trout* was overhauled for transfer to Iran, it cost $30 million—with no major modifications.

The board also pointed out that much of her machinery had been cannibalized and her instrumentation removed. Her periscope was gone. They recommended she be "stricken from the Naval Register of Ships."

On April 9, 1980, the CNO forwarded the survey to the Secretary of the Navy and advised: "The Navy has no further requirement for this ship as an operational R&D platform. It is not possible to convert this submarine to the minimum combatant capability required for modern submarine warfare. . . .

"A requirement exists for technical evaluation of a submarine launched warshot against a surfaced submarine target. Authority is requested to dispose of *Albacore* as a target to destruction for experimental purposes, as provided for in Title 10, US Code 7306. All precautions will be taken to prevent adverse environmental impact . . . sinking will be conducted more than 50 nautical miles from any coast and in water 6,000 ft. deep or greater."

The Portsmouth Submarine Memorial

Albacore was stricken from the Navy List on May 1, 1980, to await her watery grave. But her permanent resting place was not to be the depths that had several times come so close to claiming her.

A Portsmouth real estate developer with a deep interest in the seacoast area's maritime heritage, Joseph Sawtelle, had conceived and was exploring the project of a Portsmouth maritime museum. He was a leading figure in organizing the Portsmouth Marine Society, which is dedicated to publishing a series of books on the maritime history of the Maine–New Hampshire region; he himself authored a book on John Paul Jones's *Ranger*, which was outfitted and crewed from Portsmouth. He had acquired the Marconi property at Prescott Park as a likely site for the museum and was examining approaches to get the project off the ground. As he tells it, Karl Kortum, curator of the well-known San Francisco Maritime Museum, had told him, "The key to a successful maritime museum is having an elephant"—meaning a real ship.

Meanwhile, William F. "Bill" Keefe, vice mayor of Portsmouth and a city councilor, had been instrumental in bringing the Tall Ships there in the summer of 1981. Shortly afterward, a chance meeting with Russell Van Billiard at the Portsmouth Savings Bank convinced him to take on another project. Van Billiard told Keefe about seeing *Albacore* at Philadelphia, brought up his idea of a submarine museum ship, and pointed out that *569* would be the perfect choice, a testimonial to all the people who had designed her, built her, manned her. She had operated out of Portsmouth for all of her nineteen years of life. She was non-nuclear and noncombatant, eliminating possible grounds for controversy and she was the smallest sub available, simplifying problems of moving and mooring her.

Richard Winslow, in *Portsmouth Built,* quotes Bill Keefe: "I thought about that conversation that night without deciding anything. But the next morning I got a committee together to bring the *Albacore* back home."

Many people remember Bill Keefe's call. John Hart, an independent journalist and photographer, head of the Portsmouth Chamber of Commerce at the time, says, "I didn't know what he was talking about. He said, Let's bring back the *Albacore,* and I said, What the hell is an albacore?"

Keefe told Joe Sawtelle, "You should have that ship for your maritime museum." Sawtelle answered "Bill, that's a great idea"—here was the "elephant" Karl Kortum had suggested. Bob Johnston, public affairs officer at Portsmouth Naval Shipyard, remembers: "He called me and said, 'We're having a meeting. You should be there!' At that meeting we had the right combination of technical people who knew how to solve problems, and the dreamers—people who just went ahead and said, We can do it!" John Hart says, "We had about a dozen meetings over about six months, sort of fantasizing visions and stuff, until the establishment of a formal organization." Bruce Graves recalls "the original committee had petered out, but we reorganized the following year. I remember Joe Sawtelle saying to me, 'Are you with me on this? I want to follow up on it.'"

Over the winter of 1981-1982, the Bring Back the *Albacore* Committee matured into the nonprofit Portsmouth Submarine Memorial Association (PSMA), with Joe Sawtelle as chairman. Other members were Ray Hahn, commander of the Naval Reserve Unit at Portsmouth Naval Shipyard; Joe Sawtelle's business partner Harry Buron; Bruce Graves, mayor of Portsmouth; John Hallett, retired Navy man; John Clement, state commissioner of Highways and Public Works; Dick Gallant, who continued to serve on the PSMA board for many years, later managing Albacore Park; Tim Pearson, responsible for the engineering effort that eventually succeeded in bringing *Albacore* ashore; and Gene Allmendinger, then professor of naval architecture at the University of New Hampshire. And these are only a few of the many who contributed, helped, supported.

The committee had four essential challenges. First: to raise the necessary funds—which meant seeking broad public support. Second: to obtain navy consent to release the *Albacore,* which meant gaining high-level political support *and* convincing the Navy that PSMA had the organizational and fiscal ability to carry the project through. The third was to find a site; the fourth was to solve the various problems entailed in moving her and putting her on display.

Joe Sawtelle had realized early on that the Prescott Park property would not be suitable—too small, no parking. When a site-search committee was formed, it discovered that there was simply no suitable parcel of land available on the Piscataqua waterfront. But a radical and exciting solution evolved: moving the *Albacore* permanently onto dry land, completely out of the water. The idea was not totally without precedent; the captured German U-505 is mounted outside Chicago's Museum of Science and Industry. But it had never been done with a U.S. Navy memorial ship. The hurdles would be enormous. If it could be accomplished, it would be a vastly superior solution to mooring *569* permanently in the swift and powerful Piscataqua. With the river's 10-foot tidal range, *Albacore* would still be as much exposed to the forces of wind and weather as any manned and seagoing ship. Immersion in salt water would mean slow corrosion requiring constant maintenance including costly periodic dry-dockings for hull cleaning, repainting, and repair. But as Gene Allmendinger writes, "Perhaps the most advantageous aspect of an on-land site was that it would permit full viewing of the ship's unique features—its hydrodynamically streamlined hull, X-configured stern control surfaces and contra-rotating propellers." Without that hull visible, it would be impossible to *see* what made the *Albacore* unique.

Sawtelle located an ideal site, a ten-acre parcel at the main gateway to the city, at the intersection of Route 1-Bypass and Market Street. It was unused; it held the Sarah Mildred Long Bridge tollbooths, abandoned when the new Interstate 95 high-level bridge was built, but still belonged to the Maine–New Hampshire Interstate Bridge Authority. He met with Bob Allard, a trustee of the authority, who took it up with its board. The Bridge Authority board agreed this would be an appropriate use of the site, a great attraction for the community, but they were unable to make an outright donation of the land. However, the board offered to have it appraised and to sell it for its appraised value ($20,000).

Approaching the navy, the committee was told that PSMA must show it had the financial resources to complete the project before the navy would consider releasing the *569* to them. Estimates of the total cost ranged from $600,000 to $1.6 million. No public funding was available. All funds would have to be raised from private contributions. PSMA members took the idea to the residents of Portsmouth and the Maine–New Hampshire seacoast area. It seemed an overwhelming task, and some were skeptical. The *Portsmouth Herald* wrote: "Captain Ahab's battle with Moby Dick may not have been much tougher than the job Portsmouth faces in moving a black mechanical whale up the Piscataqua River." But the public's response was overwhelming; the fund drive raised almost $400,000 from the community. Russ Van Billiard said this plus the commitment of two large donors produced $758,000. Speaking in 1990, he added: "It roughly cost $1 million to do what's been done so far. We still owe $300,000—a mortgage on the land." Sawtelle says it was hoped to avoid borrowing but "people were impatient. We got letters of credit so we had the necessary cash plus the bank's guarantee." In 1998, he said, "We'll have the debt on *Albacore* paid off next year."

With a site, broad and enthusiastic public support, and assurance of adequate funding, the next step was to approach the New Hampshire and Maine congressional delegations, the Navy, and Secretary of the Navy John F. Lehman, whose support would be critical. Joe Sawtelle, Bruce Graves, and John Hallett made trips to Washington to make their case.

The Navy, however, was not so easy to deal with. Joe Sawtelle recalls: "It was delays, delays, delays. I'd call, write, and not get a direct answer. I didn't meet John Lehman until the *Albacore*'s dedication. He said, 'You're the guy who sent all the letters,' and 'I was aware of what you guys were doing.'"

Support from Congressmen was enthusiastic and readily given. Richard Winslow writes: "Congressman Norman D'Amours, representing New Hampshire's Navy Yard district, submitted bill H.R. 3980 for the sub's release (theoretically, any group from any state could have claimed the *Albacore* with a competitive offer). Following D'Amours' testimony before the House Armed Services sub-committee, both houses of Congress passed the bill. On Monday, November 7, 1983, President Ronald Reagan signed the bill for the boat's release, making front-page headlines in New Hampshire's newspapers."

The final authority whose consent was required was the Naval Sea Systems Command. Documentation supporting PSMA's ability to complete the project was provided to NAVSEA over the winter of 1983/1984. Gene Allmendinger writes, "On 20 April 1984 NAVSEA was informed that the decision had been reached to transfer *Albacore* to PSMA. However, actual acquisition of the submarine did not occur until 3 May 1985—the Hon. James F. Goodrich, Undersecretary of the Navy, formalized the transfer at a ceremony just twelve hours before *Albacore*'s final move to its site." Russ Van Billiard points out, though, that the Navy still owns the boat and could take her back in the unlikely event she is ever needed!

Bringing Her Home

Between April 20, 1984, and May 3, 1985, custody of *Albacore* was held by the U.S. Army Reserve, then for about a year by the Naval Reserve Center at Portsmouth Naval Shipyard. The plan was to move her from the Philadelphia mothball fleet to PNS, where she'd be prepared for transfer to civilian control, with the removal of classified equipment. But how? When the Navy was asked, it begged off, saying that all its available tugs were already committed; the prices private towing companies were asking were huge. The solution led to the odd result of a U.S. Navy submarine in the custody of the U.S. Army.

Luckily, Chief Warrant Officer Mark Anthony of the 69th Transportation Platoon of the U.S. Army Reserve—the Army's "tugboat unit"—happened to be visiting Portsmouth (where he later settled) and heard about the effort to get *Albacore* back home. He called Sawtelle and told him he thought it would make an excellent training mission. With Bob Johnston of PNS, they went to his bosses, got their approval, and tied up the loose ends (the Army paid). The Navy, he said, "found it a bit strange that the Army would undertake that mission, but gave it their approval if not their blessing" when they learned that Chief Anthony had been a tug skipper in the Navy before applying to the Army's warrant officer program. Thus, on April 20, 1984, NAVSEA handed custody of the *Albacore* very briefly to the Army Reserve.

Anthony inspected her at Philadelphia. He checked her for watertight integrity, that she had proper navigation lights, that her hatches were sealed. She was to make the 575-mile, 70-hour trip with no one on board; the Navy had sealed her, hatches welded closed for security reasons, because of the classified equipment she still carried aboard. She was undocked by Navy tugs 801 and 761 and handed over to the Army's big tug *Okinawa* under Chief Anthony and to the smaller LT-1953, in the Delaware River. She was on 600 feet of three-strand cable towing hawser so thick "you could almost walk on it," said Anthony.

Albacore *bucking the stiff current of the Cape Cod Canal. Joseph Sawtelle, chairman of the "Bring Back the* Albacore *Committee" (later the Portsmouth Submarine Memorial Association) and Bob Johnston of PNS are aboard the* Okinawa *(Mark Anthony)*

Almost immediately *Albacore* began to make trouble—because of her streamlining. "Once I got out of the Delaware River heading up toward Long Island," says Anthony. "we had to really step on it, she was just gliding through the water. The crew kept saying, 'I think there's somebody on board.' " The problem was that her resistance was so much less than *Okinawa*'s that whenever the tug's speed varied and she slowed down slightly, *Albacore* would shoot ahead, like she was trying to overtake her, coming up parallel with the *Okinawa*. He says *569*'s rudder had a slight turn of a degree or two, creating a slight drag: "If it had been straight ahead, she'd have run us over."

Their average speed was about 5 knots, less than Anthony had hoped; he had expected to make the Cape Cod Canal by dark, but instead they laid to off Narragansett Bay. He had originally planned to go outside Nantucket and around Cape Cod but heavy seas and northwest winds made him decide on the canal.

Albacore *under tow by the big U.S. Army tug* Okinawa *that brought her home from Philadelphia under command of chief warrant officer Mark Anthony. (Bob Johnston)*

At the entrance to the canal, Joe Sawtelle and Bob Johnston boarded. The canal gave further problems: The current was strong, really shifting. *Okinawa* shortened the towline to 150 feet and LT-1953 came alongside *Albacore* to give her a little more speed and to keep her from wandering off to the left. They hit at maximum current and were just about managing to keep ahead of it, spending two hours inside the 7.8-mile-long canal.

They arrived off the Isles of Shoals at 2 A.M. in heavy seas. To reduce rolling and strain on the tow cable, they headed seaward, northeast, for about two hours. Thus, they missed the rendezvous at Gunboat Shoal Buoy with the Moran tug that was supposed to pick up *Albacore* in time for high water at 5 A.M.. "The Moran was upset and Portsmouth was disappointed," said Anthony. "People were lining the banks of the river, it was filled with boats."

They waited for the next high water at 11 A.M., then passed the tow to the Moran. Anthony describes the docking: "Because in my experience the Piscataqua is one of the most difficult rivers to navigate except at slack water—considerable height of tide, current very swift, very nasty, shoals off in a lot of places—you have to have a docking pilot, rather than take a chance. They put a Navy crew on the *Albacore,* a line-handling party."

On Sunday, April 29, 1984, *Albacore* came home, making her last trip up the Piscataqua, to a royal welcome from her city. All Portsmouth turned out to greet her. Hundreds of boats filled the harbor, thousands of spectators lined its shores, planes and helicopters flew overhead, the air was filled with the booming of the horns of Portsmouth's tug fleet and the wail of fireboats, spraying columns of water overhead as they escorted her in. She tied up at Portsmouth Naval Shipyard's Berth 7, adjacent to the Naval Reserve Center.

Anthony recalls: "There were TV crews, newsmen, a large gathering at Pier 2 restaurant. They interviewed me and introduced the crew. They rolled out the red carpet for us. With the Navy unable to do the job, it was kind of a feather in the cap for the Army."

Beaching the Whale

The Naval Reserve Center at Portsmouth Naval Shipyard then took over *Albacore* for a year of work. It was necessary to "sanitize" the ship by removing or covering up all classified systems and items (such as the "slippery water" polymer ejection system) as directed by NAVSEA, as well as restoring to working order those necessary for her final move, including depth and speed indicators and ballast tanks. Commander Ray Hahn, CO of the Reserve Unit, recalls attending a PSMA meeting, volunteering his outfit, and contacting the commander of NAVSEA for instructions. Gene Allmendinger writes: "This extensive and complex work was accomplished by Naval Reservists during their drill periods, thus providing valuable training experience. Divers of the Reserve Mobile Diving and Salvage Unit also provided invaluable service during this time as well as at the time of the ship's landing. Needless to say, the outstanding efforts of both the Army and Navy Reservists on *Albacore* resulted in inestimable savings for PSMA and its sincere appreciation to these Reservists is recorded herein."

Bob Johnston adds, "The shipyard gave us an awful lot of support. This was the shipyard's baby, she was home-ported, repaired, modified, operated out of there. It was all Portsmouth—that's why she belongs here."

Meanwhile, Gene Allmendinger designed a system of concrete cra-

Journey's over! Albacore *off Fort Constitution entering Portsmouth Harbor. (Jean Sawtelle)*

dles to support her on dry land, under strong points of her hull structure. The forward cradle was prepared in advance: the aft one would be poured when she was in place. Both are of steel-reinforced concrete, lined with composite rubber tile, and rest on a rock ledge beneath the earth basin in which *569* rests.

Now the question was how to get her there? The Market Street site would place her about a quarter of a mile inland from the Piscataqua and about 27 feet above sea level. Plans initiated in 1983 involved consideration of at least three different methods, but the one selected was a fairly conventional "marine railway," a device familiar as a means of dry-docking small ships. The submarine would be floated onto a specially designed cradle, moving on steel rollers on a twin-rail track. This would follow a gully running through the site, reducing the substantial amount of excavation required. The mouth of the gully would be dredged to form a canal deep enough to float the submarine over the submerged cradle. A winch exerting a force of 50,000 pounds would pull the cradle out of the water—

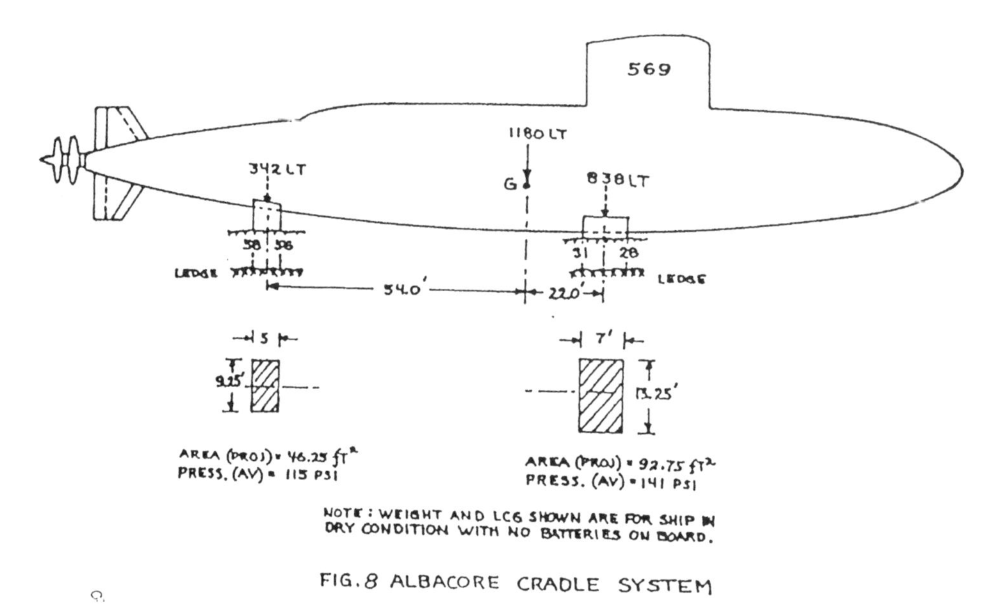

Albacore *is unusual among museum ships in that she is mounted completely out of the water on dry land. Advantages are drastically reduced maintenance and the fact that her unique hull and control surfaces are completely visible. (Allmendinger and Jackson)*

at which point it would bear the sub's total dead weight minus battery of 1,200 tons—up the ramp a distance of 650 feet in a path following the contour of the gully. The system was engineered by Crandall Drydock Co. and the cradle was built by Bath Iron Works. Gene Allmendinger says: "We went with the concept; it was a tried-and-true solution."

Two physical barriers remained: a Boston and Maine railroad trestle running over the Piscataqua some distance from the shoreline, and the brand-new Market Street Extension, a four-lane highway running between the shoreline and the site. Towing the *Albacore* through the trestle would require a thirty-foot cut. Dredging the canal to the marine railway would require a seventy-foot cut in the highway. The intensity of effort to breach and rebuild both in the shortest possible time would be enormous. And could permission to do it be obtained?

John Clement, as commissioner of highways, took the responsibility for closing the highway. Harry Buron was assigned the job of getting permission from the Boston and Maine. They agreed to shut down the railroad only with (a) written approval from all the industrial and commercial customers serviced by the dead-end line, (b) a month's

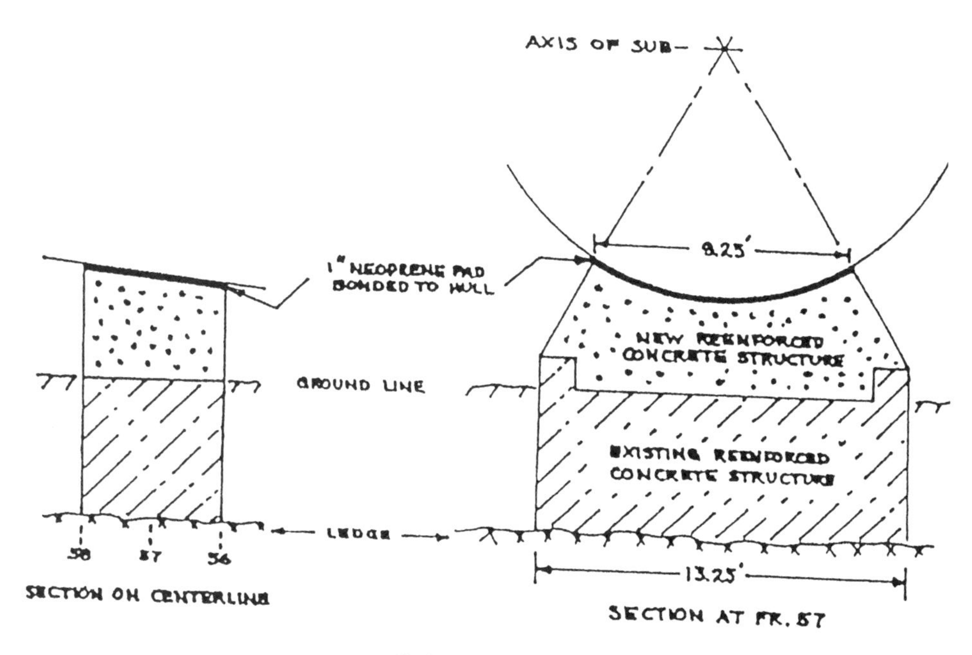

These diagrams by Professor Eugene Allmendinger show the system of concrete cradles he designed to support Albacore *on land. The cradle system seems cast-iron and foolproof; it was moving* 569 *into position on dry land that was difficult and complex. (Allmendinger and Jackson)*

notice so deliveries could be made in advance; and (c) the use of a qualified contractor they approved to tear down and put back the trestle section. All were obtained.

The organizational and human side of the challenge was truly as much effort as the engineering. To quote Gene Allmendinger: "The extended process of hurdling legal obstacles imposed by the necessity of obtaining permits and formal approval from city, state, and federal agencies and private entities began almost immediately after the site's acquisition in April of 1983. Over twenty separate permits and approvals were required for the project to proceed as well as seemingly endless hearings and defenses of its merit."

In the end, the trestle was torn down one day in advance of *Albacore*'s move and reassembled in six days. Down time for the highway was fourteen days. Ray Hahn says, "Ed MacArthur of Civil Consultants played a major role in the engineering work of making the cut in the bridge and road." But restricting an already-tight timetable

The site of Albacore *Park before the move, showing the "marine railway," the initial means planned to raise* Albacore *to her final resting place on dry land. (Russel Van Billiard)*

was the fact that the move could be made only at the absolutely highest monthly "spring" tide. The free-floating draft of the sub was 16 feet at the bow and 18.5 feet at the stern fins. To enable her to be floated over the cradle meant dredging the canal to a minimum of 17 feet below the mean sea level at that location. Floating her onto the cradle would still have to be achieved during a very narrow window of time, a fact that would lead to major problems.

On May 4, 1985, began the process of "beaching the whale"; if all went as planned, she would be in position at her final resting place by the end of the day. She was towed without incident across the Piscataqua, with about twenty former crew members on board to operate her fully functioning ballast tank blow system to control her buoyancy. Spectators lined the shore and bridges. At 9:45 A.M. she passed stern-first through the trestle, where three shallow draft work boats, a "spud barge" with winches, heavy earth-moving equipment, and numerous volunteer line-handlers were waiting to take over.

Her path between the trestle and highway cut formed a dogleg.

She had to be moved several hundred feet sideways to the right, with first her stern swung into position, then her bow pulled starboard to line her up with the highway cut. But at this point, her stern swung to *port*, and the lower fins of her X-stern went aground in the mud. She stuck fast; they had great trouble breaking her loose as high tide came and went.

The former crew members on board got her off by blowing ballast tanks. Through the cut and centered precisely over the cradle at 12:30, with only minutes to spare of the high tide "window of opportunity," disaster struck. Her propellers had been removed, and the long projecting prop shaft had been covered with a protective structure, a "tin can" or "dunce cap." Its presence had not been considered by the designers of the marine railway; *Albacore*'s stern would not mate with the cradle. The strength of the cradle's structural members had been carefully calculated to bear the huge weight of the submarine, but it was impossible to line her up with the precise reference points. Still, there was no choice but to go ahead. Ray Hahn says: "We thought we were close—we thought we were safe. The tide was going out—the weight seemed okay—we tried the winch. It totally failed to work. The problem turned out to be incorrect lubricant—we fixed this.

"The next day we started the winch. She began to move a foot, 10, 25, 30 feet. It was when her total weight was out of the water that the cradle failed. The beams just buckled and it slipped off the tracks. Had it stayed on the track, the move could possibly be completed.

"We left it on the cradle for a while. Then we floated the ship off, using the air and ballast tank systems, still workable today. We moved her back as far as possible at high tide, let the ship settle as the tide went out, let nature take its course."

And this was very likely the absolute low point in the entire history of the *Albacore*. Gene Allmendinger says, "If the marine railway had worked, she would have been in place the same day. Instead she lay a forlorn mass of steel for about six months. This was a tremendous disappointment. It was a financial shock as well.

Less than appreciated by those who had worked endless hours on this projects was the media's jocular reference to the *Albacore* as the "albatross."

Ray Hahn recalls "There was no contingency plan. We couldn't remove the ship without closing the bridge and railroad again. The only other solution was to cut the ship up, and, well, nobody was about to do that!"

On May 4, 1985, Albacore *was towed without incident from the shipyard. Here the tug has her stern lined up with the cut in the Boston and Maine Railroad trestle. The cut in the Market St. Extension and canal dredged to her display site are visible. (Portsmouth Herald and Norman Bowers)*

Vessels of various sizes, heavy equipment, and and a lot of strong arms and legs were used to position Albacore *for the haulout. Spectators were treated to a once in a lifetime event. (Top photograph by Peter E. Randall, others Portsmouth Herald)*

Albacore *being aligned on its cradle. Neglecting to remove the oversized sheath on the propeller shaft prevented proper alignment over the cradle causing the cradle to collapse. (Jean Sawtelle)*

Bob Johnston expresses the atmosphere of these trying times: "For six months she lay on her side in the mud. I took to driving another way to work so I wouldn't see her. It was such a wonderful project, then it went awry."

But as Bob says, some of the board "saw the nobility of the project" and were determined to see it succeed: Joe Sawtelle, Gene Allmendinger, Bruce Graves, Tim Pearson, among others. Gene Allmendinger says: "All of us contributed. Sawtelle clearly emerged as the one to see the project through to completion—he was not one to fail to assume responsibility. I was coordinator of the technical aspects."

They reviewed possible engineering solutions very carefully and ultimately settled on a method of building a huge earthwork, a coffer-dam or "bathtub" around the ship that would be flooded with water until the submarine floated free. A series of new, higher-level walls would be built to form a succession of "canal locks." With the flooding of each new "lock." *Albacore* would be moved to a position farther inland and at a higher elevation.

Tim Pearson provided the direction for construction of the "bathtub." No simple task; Gene Allmendinger says: "It was a great problem of soil mechanics. We had to establish a hydrostatic head 6 1/2 feet above the Route 1 bypass; you could see what would happen if we had a blowout and that volume of water broke loose." The University of New Hampshire School of Engineering provided technical assistance; before granting permission for the project to go ahead, the Army Corps of Engineers required a thorough analysis of the soil mechanics and geometry of the structure as well as the system of lining its walls with poly sheet to make them as leakproof as possible.

"If I gave one thing to the project," says Bruce Graves, "it was that I took it through the permit process with the Army Engineers. I hand-delivered every document. The people of the corps were kind of incredulous."

Sawtelle says, "The next problem was, how to fill it with water? We searched all over New England for pumps big enough; our original pumps just couldn't produce the necessary volume of water. Warren Pratt, president of Walter S. Pratt and Sons, Inc., of Rensselaer New York, had 12 huge pumps that he donated—never charged a penny. He hauled them all the way from New York at his own expense.

"And the next problem was that the pumps would run out of gas every three hours! It took a huge amount of gas—they were running 24 hours a day. Buzzy Hanscom, who owns a fuel company, came by with his tanker truck, and filled the tanks, kept filled all night–and he never charged us a penny."

Finally, all details were complete. Three successive floods and floats were required. As Gene Allmendinger describes it, "In a three-day, around-the-clock operation, workmen flooded the tub with 9.5 million gallons of water to a height of six feet above the adjacent highway. The ship was towed the final 250 feet and, with the water level falling, settled perfectly on its cradle at 4:30 P.M. on 3 October 1985. *Albacore* was finally landed."

Triumph!

The Final Effort

Still, many months of work—including some major operations—remained to be done before *Albacore* would be ready for the public. Mounting her twin contra-rotating props required cranes and a large volunteer crew. The Navy's permission was needed to cut two visitor

Albacore *in cofferdam, top, before filling with water, and below, in final position. Note the exposed plastic liner. (top photograph by Craig Blouin, bottom by Jean Sawtelle)*

access doors in the hull of the ship, the entrance forward and exit aft. Brows or gangways leading to these doors were loaned by the shipyard, bridging the gap over the pit in which *Albacore* sits. The hull was completely sandblasted and painted in two weeks, but cleaning up her interior stretched out over the year and was not completed until August 29, 1986, the day before she was opened to the public! Standard external shore power was brought in and new wiring to handle it was installed, for service needs such as lighting, heaters, blowers, and various ship circuits, including alarm systems and indicator lights to enhance the sense of realism. A refurbished, working periscope was provided by the Navy and installed. The 1MC internal communications system was activated and a telephone was installed. City safety regulations required fire extinguishers, battery-powered emergency lighting, and bull horns.

All of this work was done by volunteers, including many shipyard workers from Portsmouth Naval Shipyard. Roy Syphers recalls: "We had quite a large crew—machinists, riggers, crane operators. It depended upon the project; for mounting the props we had a pretty large crew, but other times it was just a couple of us working on Saturdays.

"Russ Van Billiard was instrumental in coordinating it all. He'd call me up in the middle of the night and say, 'Can you get over and do this?' He coordinated us, got tools and equipment for us, served as liaison with the shipyard.

"We got a lot of support from the Navy. They provided us with all the official plans and granted us permission to use the material. It was very gratifying—people were very helpful."

Roy talks about the importance of the shipyard to the Portsmouth area—"everyone either has a relative or knows someone at the shipyard"—and the importance of *Albacore* as a monument to their skills and work. "People in my age group are very proud and attached to it," he says. "The fact is that submarine work must be of high quality, absolutely dependable. It becomes inbred after a while. It's a whole different construction trade, working in the various tanks and confined areas with inadequate lighting and little space. I remember my first experience working inside a submarine's tanks; I got hung up by my pants on a couple of pipe hangers. I had to be talked out of there—which included leaving my pants behind!"

He recalls other volunteers: Ray Hahn, Bruce Graves, Frank Bickmore ("he was instrumental in mounting the screws"), Barbara

Many months of work went into preparing Albacore *for visitors, including cutting two access doors in her hull, and placing gangways to bridge the gap between* 569 *and the wall of her earth "drydock." The present visitor's center building is visible in the background. (Largess)*

Voisine ("a sheet metal worker, she did a lot of work in the interior"), Charlie Lawrence ("still foreman in the machine shops at the shipyard"), and Vaughan Smith ("our electrical foreman").

Bob Fisher recalls: "I did the electrical wiring conversion from naval ship to civilian shore wiring, three phase to single phase. Lots of wires were not on the plans after years of retrofitting, and had to be traced out by hand."

Roy Syphers says: "We had some fun, donated some of our free time, some weekends. It's really great to see her sitting there. They still get pretty good Navy yard cooperation. Crews of ships go over and paint her." The volunteer help is thus still an ongoing thing. Hahn explains: "She's about due for another paint job. Here's how we do it. A ship will be in at the yard, we'll ask the enlisted crew to volunteer, the shipyard will donate over-shelf-life paint, we'll have a barbecue for the guys afterward. It's a goodwill effort. *Trepang* and *Helena* have done it. She needs a new paint job about every five years."

Three individuals instrumental in saving the Albacore: *Joseph Sawtelle, first president of the Portsmouth Submarine Memorial Association, former Secretary of the Navy John Lehman, and former Portsmouth City Councilman William Keefe, organizer of the first informal "Save the* Albacore *Committee."*

Early sketches had envisioned a two-story Port of Portsmouth Maritime Museum sharing the site with *Albacore*, and containing extensive exhibits, conference rooms, and an auditorium. For the moment, this dream could not be fulfilled; but a smaller visitors center was constructed. It would serve for selling tickets, souvenirs, and books, displaying submarine memorabilia, and showing a slide-and-tape show on *569* created by Gene Allmendinger.

Once the parking lots and access roads were completed, a memorial garden, a secluded place for reflection and remembrance was designed and created by Jean Sawtelle. Inside an oval of tall evergreens are a reflecting pool and granite monuments with seasonal plantings, and a black Canadian granite carving of a dolphin by Maine artist Cabot Lyford. A symbol of the submarine service, the dolphin is also a symbol of immortality.

Monuments commemorate the loss of *Squalus*, *Thresher,* and the *0-9,* a World War I sub reactivated as World War II approached and lost

Shipyard worker Barbara Voisine, one of the many volunteers who spent countless hours working to prepare Albacore *for opening to the public over roughly a year following her successful move to her permanent display site on land.*

off the Isles of Shoals while training a crew June 20, 1941. Bronze plaques bear the names of all those who died aboard these subs. A fourth monument is dedicated to the fifty-two submarines "still on patrol," with 374 officers and 3,131 men aboard who did not return from World War II. The first *Albacore,* with Jim Blanchard and Chief Stanton still aboard, heads the list.

As the Athenian Pericles put it at another memorial service long ago, "The whole earth is the tomb of heroic men, and their story is not graven only on stone over their clay, but abides everywhere without symbol, woven into the stuff of other men's lives."

On August 30, 1986, Albacore Park was opened.

Since that day, *Albacore* has had 30,000 to 50,000 visitors per year. At the Memorial Garden, a ceremony is held every Memorial Day; and submarine veterans commemorate the loss of the *Thresher* there every April 10. Major events have included *Albacore*'s designation as a National Historic Landmark by the Department of the Interior in 1989; Lando Zech was the main speaker at the ceremony. *Albacore*'s

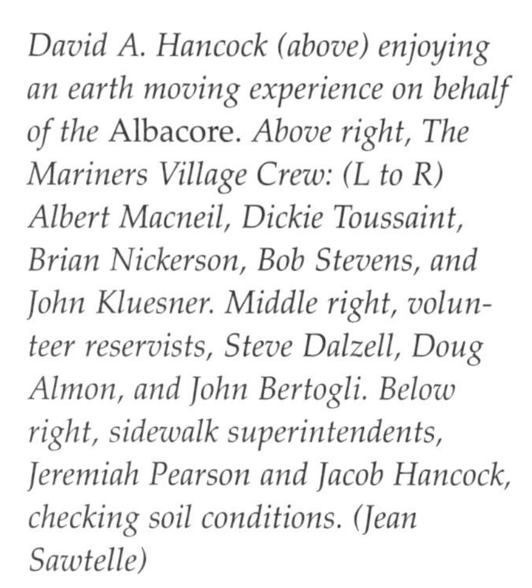

David A. Hancock (above) enjoying an earth moving experience on behalf of the Albacore. *Above right, The Mariners Village Crew: (L to R) Albert Macneil, Dickie Toussaint, Brian Nickerson, Bob Stevens, and John Kluesner. Middle right, volunteer reservists, Steve Dalzell, Doug Almon, and John Bertogli. Below right, sidewalk superintendents, Jeremiah Pearson and Jacob Hancock, checking soil conditions. (Jean Sawtelle)*

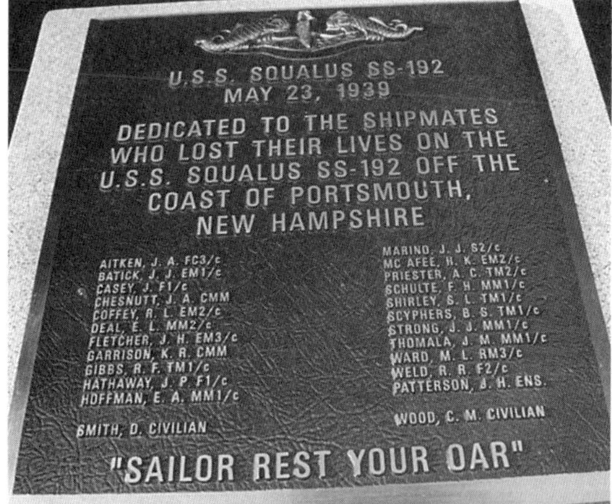

Today's Albacore *Park includes a memorial garden dedicated to those US submarines "Still on Patrol" - which failed to return in World War II. The first* Albacore, *with Jim Blanchard and Chief Stanton aboard, heads the list. (Norman Bower)*

veterans held their 39th reunion there; Ken Gummerson was the speaker, sadly, only a short time before his death. Numerous World War II submarines have had their reunions there, and frequent reenlistment ceremonies have been held for sailors from the shipyard. "We've even had weddings there, two to my knowledge," says Russ Van

Silver service was presented by Mrs. Jowers to Albacore *when the submarine was launched. It is now part of the collection of Port of Portsmouth Maritime Museum.*

Billiard. "One was a couple of civilians just looking for an interesting place to get married. But one was a Navy man. It was held topside, on the deck. Good thing that they didn't fall off!

Jim Sergeant, park manager since 1994 and a former SSBN submariner him self, notes "We've had the Scouts aboard for numerous sleepovers. We had a major camporee over at the shipyard and took several hundred Scouts through the boat that weekend. We have school groups, church groups, all kinds of organizations from all over New Hampshire and Maine. We just had a group of foreign exchange students from Jamaica. Things like that are just going on forever. You never know when that phone rings who's going to be calling."

The Port of Portsmouth Maritime Museum

And what lies in the future? According to Joe Sawtelle, the long contemplated Port of Portsmouth Maritime Museum is on the verge of becoming reality. In fact, the Portsmouth Submarine Memorial Association has already changed its name to Port of Portsmouth Maritime Museum, reflecting the board of directors' attitude: "Now we've got the sub, let's get the museum built."

The first priority was the retirement of the debt for *Albacore,* totaling $500,000 at its maximum. This should be paid off by fall 2000. Admissions have covered most of it, paying off $25,000 of the principal, plus the interest, each year. In July 1998, Sawtelle said: "We're in the design and planning stage right now—our goal is to break ground in the next few years. "

Will the museum resemble the original plan? Possibly not. Sawtelle says he has gone to a second firm of architects for a new design—a laminated beam, or "glulam," structure. "It looks like a ship house," he says. "It uses all exposed beams, and gives this sense of tremendous spaciousness inside. The North Carolina Maritime Museum uses this type of design, but it's different from anything that's been built in this region. It will have three levels, large exhibits on the lower level, an atrium on the first floor, a mezzanine on the intermediate level. The question now is, Can we finance that? The preliminary plans will cost us $20,000 for an engineer to do the structural analysis. Should we design a building we may not be able to finance? Or stick to something simpler, conventional, and much more feasible for maybe half the cost? I say, let's see if we can do the optimal plan. Let's design the building, price it out, and see what happens."

Sawtelle explained what "Port of Portsmouth" means. "It goes back to the Revolutionary period. Many of the surrounding Maine and New Hampshire towns, Kittery, Berwick, Dover, Newmarket, Exeter, had shipyards in the days of sail. Of course they're gone now—the rivers are silted up.

"The *Nightingale,* one of the most famous clippers, was built in the Hanscom Shipyard. Ten percent of the clippers, twenty-eight ships, were built here, along with hundreds of other ships from the early 1800s on. And naval history joins Portsmouth maritime history with the establishment of the Portsmouth Naval Shipyard in 1800. It's a great story."

And the *Albacore,* of course, is ultimately only a chapter in this great story, the story of Portsmouth shipbuilders and sailors. It unfolds without a break from colonial days, to John Paul Jones's *Ranger,* to the primitive "pig boats" of the early twentieth century, to World War II, to nuclear power. The ingenuity, courage, vision, and leadership necessary to create the *Albacore* were there from the beginning, part of the American seafaring tradition. With a little more of those same ingredients, the Port of Portsmouth Maritime Museum will soon become a reality.

References

The authors apologize for the lack of footnotes. The attempt has been made to cite all books and individuals directly quoted in the text itself. The main sources used are summarized below.

Introduction: Material on the history of Portsmouth and the Portsmouth Naval Shipyard is from Richard Winslow's *Portsmouth -Built.*

Chapter 1: Summary of U-boat campaigns in WW I and II as well as the development of the Type XXI is largely based on Tarrant's *The U-Boat Offensive 1914-1945.* Postwar American reaction to the Type XXI is described in Gary Weir's *Forged in War* and Friedman's *Submarine Design and Development* and *US Submarines Since 1945.* The U.S. submarine campaign against Japan is largely based on Clay Blair's *Silent Victory.* The account of the Battle of the Philippine Sea is based on Blair, Morison's *Two-Ocean War,* and Nomura's "Ozawa in the Pacific, a Junior Officers Experience" in Evan's *The Japanese Navy in WW II.* The story of the "first" *Albacore* is based on her history in Mooney's *Dictionary of American Naval Fighting Ships*, Blair, and Ned Beach's *Submarine! Albacore's* launching is taken from accounts in the PNS newspaper *Periscope* and interviews with Mrs. Kraus.

Chapter II: Information on Adm. Momsen is from Maas' *The Rescuers*, his speeches and papers at the Naval Historical Center's Operational Archives, and interviews with Adm. C. B. Momsen, Jr. Information on the CUW is based on NAS and its "Naval Studies Board" archives and interviews with Frank Andrews, John Coleman and George Wood, Dan Sawitsky. Michael Bruno provided documents and interviews on Kenneth Davidson and the role of Stevens Institute. Also used were the Louis Landweber oral history at Navy Labs Archives at Whiteoaks, Md., and the John Niedermair oral history, minutes of the Submarine Officers Conference and documents of the Ships' Characteristics Board all held at the NHC Operational Archives. Early design decisions are also described by Harry Jackson in interviews and "*Albacore, Past-Present-Future,*" and *Forged in War.* Discussions with Dr. Gary Weir provided the direction for this research.

Chapter III: Based on Harry Jackson and Gene Allmendinger interviews and articles, Friedman's *Us Submarines Since 1945* and Bureau of Ships correspondence at the National Archives. The *Spruce Goose* info is from Knott's *History of the American Flying Boat.* Much of the story of *Albacore's* completion and trials is from an interview of Ken Gummerson by Harvey Horowitz, and his speech at *Albacore's* 39th reunion.

Chapter IV: This chapter is mostly based on interviews, and correspondence with Jon Boyes, Ted Davis, and Dr. Henry Payne, as well as Boyes' "Flying the Albacore" *Submarine Review* article, Davis "Will our Subs Have a Fighting Chance" in the USNI *Proceedings,* and Dr. Payne's numerous articles in both journals. The Mountbatten/Rickover relationship is described in Grove's *Vanguard to Trident.* Thanks to Richard Van Treuren and Althoff's *Sky Ship* for info on "The Flying Wind Tunnel." See Engelhardt's "Soviet Sub Design Philosophy" for a succinct summation of points found in many works.

Chapter V: Background on Admiral Rickover is based on Hewlett and Duncan's *Nuclear Navy*, biographies by Duncan and Polmar and Allen, and numerous interviews, books, and articles by those who had contact with him, including critics such as Harold Hemond. The origin of the *Skipjack* is from *Forged in War.* The general chronology of *Albacore's* modifications and movement is from her ship's history, obtained at the NHC. Other information is from interviews with Ned Beach (subject, the *Triton*) Bill Day, Dick Stensen and Jerry Feldman of DTMB, Ted Davis, Harry Jackson, Lou Urbanczyck, Lando Zech, Jon Boyes, and Howell Russell.

Chapter VI: Background to the *Los Angeles* design can be found in Friedman's *US Submarines Since 1945,* also Duncan's *Rickover,* with criticism from Tyler's *Running Critical*. Bill St. Lawrence is the main source for the x-stern and dive brakes, also Ron Hines,

John McCarthy, and Howell Russell. The origins of SOSUS is from *Forged in War.* DIMUS is from Frank Andrews, Bill St. Lawrence, and *US Subs Since 1945.* Towed sonar is from Friedman's books and Marvin Lasky.

Chapter VII: Incorporates much chronology and detail from *Albacore's* Ship's History. Contraprops and the silver-zinc battery are from Friedman and interviews with Capt. Roy Springer, Stan Zajechowski, and Russ Van Billiard. AUTEC is described in Bentley's *The Thresher Disaster* and by Marvin Lasky, the *Dolphin* and MONOB I in Polmar's *Ships and Aircraft of the US Fleet.* The FAB is described by Howell Russell and Marvin Lasky. David Kratch's report on Project SURPASS is held by the NHC's Ship's History Division.

Chapter VIII: Adm. Grenfell's memo. Robert Ball's summary of 569's material condition, the 1979 survey, and the CNO's recommendation for the disposal of *Albacore* as a target are all from *Albacore's* folder in the NHC's Ships' History Division. The story of *Albacore's* preservation is from Richard Winslow, the many interviews cited, and Allmendinger and Jackson's "*Albacore* Past-Present-Future."

ARCHIVAL SOURCES

The main archival sources, including photo archives, consulted are:

— Navy Labs Archives, formerly at David Taylor, later at the Naval Surface Warfare Center at Whiteoak, Md.
— Naval Historical Center Operational Archives Photo Archives, and Ship's History Division.
— National Archives Bureau of Ships Correspondence, and Still Pictures Division.
— National Academy of Sciences Archives
— NAS Naval Studies Board Archives
— US Naval Institute Photographic Collection
— Portsmouth Submarine Memorial holdings
— Portsmouth Naval Shipyard Museum

UNPUBLISHED DOCUMENTS

Allmendinger, Eugene, and Capt. Harry Jackson. "*Albacore* Past-Present-Future," paper presented to the New England Section of the Society of Naval Architects and Marine Engineers, May 6, 1989.

Bruno, Michael S., Director, Davidson Lab. "Davidson Laboratory and the Experiments of Towing Tanks; The History of Towing Tanks Research at Stevens."

Gertler, Morton. Head, Stability and Control Division David Taylor Model Basin. "*Albacore* Research Program."

— "Potentialities of the *Albacore* -Type Submarine." Paper presented to the American Society of Mechanical Engineers, Philadelphia, PA., 12 November, 1958.

"Interim Report of the Panel on the Hydrodynamics of Submerged Bodies, 7, November, 1949." Naval Studies Board Archives, National Academy of Sciences, Washington DC.

"Landweber, Louis: Oral History." Interview by David K. Allison 22, January, 1986. Navy Laboratories Archives, Naval Surface Warfare Center, Whiteoaks, MD.

"Minutes: Seventh Meeting of the Panel on the Hydrodynamics of Submerged Bodies. Underwater Warfare Committee, National Research Council; 26 October, 1949."

Momsen, R. Adm. Charles B., "Launching of the *Albacore*" address by Adm. Momsen, Portsmouth Naval Shipyard 1 August, 1953.

— "Remarks by R. Adm. C. B. Momsen USN at Commissioning USS *Albacore* Portsmouth, NH. 5 December, 1953."

Wislicenus, Dr. G. F. "Report on Observations and Conclusions Concerning Hydrodynamics of Submerged Bodies and Related Subjects. Submitted to the panel on Hydrodynamics of Submerged Bodies Undersea Warfare Committee. National Research Council." 1949?

PAMPHLETS

"Committee on Undersea Warfare" Committee on Undersea Warfare, National Academy of Sciences, National Research Council, Washington, DC, 1971.

"Decommissioning Ceremony - USS *Albacore* (AG55-569) 1, September, 1972."

"Information Booklet: The David Taylor Model Basin; Authorized By Act of Congress approved May 6, 1936; Dedicated November, 4, 1939."

"Towing Tanks and Hulls" Hoboken: Stevens Institute of Technology, 1937.

"USS *Albacore* (AG55-569) Commissioning Ceremonies. Saturday, 5 December, 1953." Portsmouth, NH: Portsmouth Naval Shipyard 1953.

ARTICLES

Ashley, Richard T. "Remembering the Diesel." *US Naval Institute Proceedings,* Volume 115/7/1,037, July 1989. p 96-100.

Andrew, Frank A. "Antisubmarine Warfare." *The Encyclopedia of Physical Science and Technology,* Vol. 1, p 644-660. The Academic Press, 1987.

Beach, Capt. Edward L. "Pictorial—US Nuclear Powered Submarines." *USNI Proceedings* Vol. 93 no. 8 whole no. 774, August, 1967. p 87-101.

Boyes, Vice Adm. Jon L. "Flying the *Albacore.*" *The Submarine Review,* July 1988. p 14-22.

Broad, W. J. "Navy Has Long Had Secret Subs For Deep-Sea Spying, Experts Say." *The New York Times*, 7, February, 1994. pp A-1 and B-7.

Brown, D. K. Review of A. Compton-Hall, *Submarine Versus Submarine* in *Warship 1989.* London: Conway Maritime Press 1989 p 205-206.

Corlett, Cmdr. Roy, RN. "Soviet Submarine Propulsion: Signs of a Great

Leap Forward?" *Jane's Naval Review,* London: Jane's Publishing Co., Ltd., 1986.

Davis, Capt. Theodore F. "Will Our Subs Have a Fighting Chance?" *USNI Proceedings,* Volume 114/8/1026, August 1988. p 107-109

Davidson, Kenneth S. M. "The 1947 Report of the American Towing Tank Conference." *Transactions Vol. 55* New York: The Society of Naval Architects and Marine Engineers, 1947. p 492-494

— "The 1948 Report of the American Towing Tank Conference." *Transactions Vol. 56*. New York: The Society of Naval Architects and Marine Engineers, 1948 p 571-575.

Englehardt, John J. "Soviet Sub Design Philosophy." *USNI Proceedings,* Volume 113/10/1016. October 1987 p 193-200.

Friedman, Norman. "US ASW SSK Submarines." *Warship.* Vol. VIII. London: Conway Maritime Press, 1984. p 92-99.

Gruner, William P. and Henry Payne III. "Submarine Maneuver Control." *USNI Proceedings,* Volume 118/7/1073, July 1992. p. 56-60

Hart, Ken. "Submarine Automation." *The Submarine Review,* July 1988 p 49.

Head, Brian "The Hawkcraig Experiments; The Beginnings of Submarine Detection." *Warship,* Volume 49, January 1989. p 7-16.

Hemond, Harold C. "The Flip Side of Rickover." *USNI Proceedings,* Volume 115/7/1037, July 1989 p 42-47.

Jackson, Capt. Harry, William D. Needham, and Dale E. Sigman. "ASW: Revolution or Evolution." *USNI Proceedings,* Volume 112 September 1986 p 64-71.

Largess, Robert P. "USS *Triton*: The Ultimate Submersible *The Submarine Review,* January 1994 p 101-107.

— and Harvery S. Horwitz "*Albacore*—the Shape of the Future." *Warship 1991*, ed. Robert Gardiner, London: Conway Maritime Press. 1991.

— "*USS Triton:* the Ultimate Submersible." *Warship 1993.* London: Conway Maritime Press, 1993.

Macintyre, Capt. Donald, RN. "Shipborne Radar." *USNI Proceedings,* Vol. 93, no. 9, whole no. 775, Sept. 1967. p. 70-83.

McKee, Andrew A. "Recent Submarine Design Practices and Problems" *Transactions Vol. 67.* New York: The Soc. of Nas. Architects and Marine Engineers, 1959. p 623-652.

Merriman, David D. "*USS Albacore;* Forerunner of Today's Submarines." *Sea Classics,* Vol. 80, January, 1980 p 18-29.

Miller, David. "The First Hunter-Killers: British 'R' Class Submarines of 1917." *Warship 1993.* London: Conway Maritime Press, 1993. p 167-187.

Nylen, Lt. David I. "Melee Warfare." *USNI Proceedings,* Volume 113/10/1016, October 1987. p 56-64.

Payne, Henry E III, Frank R. Carter, and Wayne J. Harrison. "The Navy's Flying Wind Tunnel." *Aerospace Engineering,* Vol. 20, no. 2, March 1961. p 18-19, 50-61.

Payne, Henry E III. "The *Albacore*: Back to the Future." *USNI Proceedings,* Volume 119/7/1085, July 1993. p 59-62.

— "A Keel Wing to Solve the Snap Roll." *The Submarine Review,*. Date? p 3-11.

— "Our Subs Fly With One-Half a Wing." *The Submarine Review.* Date? p 28-34.

— "Want to Study About Modern Submarine Design?" *The Submarine Review.* Date? p 73-79.

— "An Airborne Underseas Weapon System." *The Submarine Review.* February 1963. p 51-62.

— "Submarine Maneuvering Instability." *The Submarine Review.* Date? p 48-59.

The Portsmouth Periscope (later: *The Periscope, Official Newspaper of the Portsmouth Naval Shipyard*) 3/12/52, 7/31/53, 8/14/53, 12/5/54, 4/1/55, 6/24/55, 7/8/55, 10/7/55, 1/6/56, 1/24/56, 6/29/56, 1/24/58, 7/18/58, 3/29/59, 7/22/60, 3/31/61, 11/3/61, 4/16/71, 7/12/62, 7/17/64, 1/14/66, 1/28/66, 3/25/66, 5/20/66, 9/23/66, 10/7/66, 10/21/66, 11/4/66, 12/2/66, 12/16/66, 12/30/66, 1/13/67.

Perlow, N. K. "My Terrifying Dive in the World's Fastest Submarine." *The Police Gazette,* 1955.

Polmar, Norman, and D. A. Paolucci. "What Killed the Scorpion?" *Seapower,* May 1978. p13-19.

Rickover, R. Adm. H. G., Capt. J. M. Dunford, Theodore Rockwell, Lt. Cdr. W. C. Barnes, and Milton Shaw. "Some Problems in the Application of Nuclear Propulsion to Naval Vessels." *Transactions Vol. 65,* New York, The Society of Naval Architects and Marine Engineers. 1957. p 714-736.

Ryan, Cornelius. "I Rode The World's Fastest Sub." *Collier's,* 135, April 1, 1955. p 25-29.

Schade, Commodore Henry A. "German Wartime Technical Developments." *Transactions Vol. 54.* New York: The Society of Navyal Architects and Marine Engineers. 1946. p 85-111.

"Telltale Tank." *Newsweek,* October 17, 1955. p 114-115.

BOOKS

Albrecht, G. *Weyer's Warships of the World,* 57th Edition 1984/5. Annapolis: The Nautical/Aviation Publishing Co. of America, 1983.

Alden, John D. *The Fleet Submarine in the US Navy; a Design and Construction History.* Annapolis: Naval Institute Press, 1979.

Althoff, William F. *Sky Ships; A History of the Airship in the United States Navy.* New York: Orion Books, 1990.

Anderson, Comdr. William R. *Nautilus 90 North.* Cleveland: World Publishing Co., 1959.

Beach, Capt. Edward L. *Around the World Submerged; the Voyage of the Triton*. New York: Holt, Rinehart, and Winston, 1962.

—*Cold is the Sea*. New York: Holt, Rinehart, and Winston, 1978.

—*Submarine!* New York: Holt, Rinehart, and Winston, 1946.

Bekker, C. D. *Defeat at Sea: The Struggle and Eventual Destruction of the German Navy 1939-1945*. New York: Ballantine Books, 1956.

Bentley, John. *The Thresher Disaster: The Most Tragic Dive in Submarine History*. Garden City: Doubleday and Co, 1974.

Blair, Clay. *Silent Victory: the US Submarine War Against Japan*. Philadelphia and New York: J. B. Lippincott, 1975

Blake, Bernard, ed. *Jane's Underwater Warfare Systems 1990-1991*. Alexandria: Jane's Information Group, 1990

Brown, D. K. *The Future British Surface Fleet: Options for Medium-Sized Navies*. London: Conway Maritime Press, 1991

Calvert, Comdr. James. *Surface at the Pole*. New York: McGraw-Hill, 1960

Churchill, Sir Winston S. *The Second World War*, 6 Vols. London: Cassell, 1952.

Cocker, M. P. *Observer's Directory of Royal Naval Submarines 1901-1982*. London: Fredrick Warne, 1982

Compton-Hall, Richard. *Submarine Versus Submarine: The Tactics and Technology of Underwater Warfare*. New York: Orion Books, 1988

—*Submarine Warfare: Monsters and Midgets*. Blandford Press, 1985.

Dodgson, The Rev. Charles L. *The Humorous Verse of Lewis Carroll*. New York: Dover, 1960

Duncan, Francis. *Rickover and the Nuclear Navy; The Discipline to Technology*. Annapolis: Naval Institute Press, 1990.

Dunham, Roger C. *Spy Sub.* Annapolis: Naval Institute Press, 1996.

Enright, Capt. Joseph F., and James W. Ryan *Shinano! The Sinking of Japan's Secret Supership.* New York: St. Martin's Press, 1987.

Evans, David C. ed. *The Japanese Navy in World War II: In the Words of Former Japanese Officers.* Annapolis: Naval Institute Press, 1986.

Fahey, James C. *The Ships and Aircraft of the US Fleet*, Victory Edition. New York: Ships and Aircraft, 1945.

—*The Ships and Aircraft of the US Fleet, Seventh Edition.* Falls Church: Ships and Aircraft, 1958.

Farrago, Ladislas. *The Tenth Fleet.* New York: Obolensky, 1962.

Friedman, Norman. *The Postwar Naval Revolution.* Annapolis: Naval Institute Press, 1986.

—*The Naval Institute Guide to World Naval Weapons Systems,* 1991/1992. Annapolis: Naval Institute Press, 1991.

—*Submarine Design and Development.* Annapolis: Naval Institute Press, 1984.

—*The US Maritime Strategy.* Annapolis: Naval Institute Press, 1988.

—*US Naval Weapons.* London: Conway Maritime Press, 1983.

Gabler, Ulrich. *Submarine Design.* Koblenz: Bernard and Graefe Verlag, 1986.

Gorshkov, Adm. Sergei G. *The Sea Power of the State.* Annapolis: Naval Institute Press, 1976.

Gray, Edwyn *The Devil's Device.* Annapolis: Naval Institute Press, 1991.

Grove, Eric J. *Vanguard to Trident: British Naval Policy Since World War II.* Annapolis: Naval Institute Press, 1987.

Hackmann, Willem. *Seek and Strike: Sonar Anti-Submarine Warfare, and the Royal Navy* 1914-1954. London: Her Majesty's Stationary Office, 1984.

Herken, George. *The Winning Weapon: The Atomic Bomb in the Cold War, 1945-1950*. Princeton: Princeton University Press, 1981.

Herrick, Robert W. *Soviet Naval Theory and Policy; Gorshkov's Inheritance*. Newport: Naval War College Press, 1988.

Hewlett, Richard G, and Francis Duncan. *Nuclear Navy, 1946-1962*. Chicago: University of Chicago Press, 1974.

Hezlet, Adm. A. R. *The Electron and Sea Power*. London: Peter Davies, 1975.

Hill, R. Adm. J. R. *Anti-Submarine Warfare*. Annapolis: Naval Institute Press, 1985.

Jentschura, Hensgeorg, Dieter Jung, and Peter Mickel. *Warships of the Imperial Japanese Navy 1869-1945*. Annapolis: Naval Institute Press, 1986.

The Joint Army-Navy Assessment Committee, *Japanese Naval and Merchant Shipping Losses During World War II by All Causes*. Washington: Govt. Printing Office, 1947.

Kahn, Herman. OM Thermonuclear War. Princeton: Princeton University Press, 1960.

Knott, Richard C. *The American Flying Boat; an Illustrated History*. Annapolis: Naval Institute Press, 1979.

Kuenne, Robert E. *The Attack Submarine; a Study in Strategy*. New Haven and London: Yale University Press, 1965.

Levine, Robert. *The Arms Debate*. Cambridge: Harvard University Press, 1963.

Maas, Peter. *The Rescuer*. New York: Harper and Row, 1967.

Macintyre, Capt. Donald. *U-Boat Killer.* Weidenfeld and Nicholson, 1956.

Mandelblatt, James L. *Rebirth of a Submarine; A History of the USS Requin (SS-481 / SSR-481),* 1995.

Miller, David. *Modern Submarines.* London: Salamander Books, 1989.

Mooney, James L. and Richard T. Speer, eds. *Dictionary of American Naval Fighting Ships 8 Vols.* Washington: GPO, 1959-1992.

Moore, Capt. John E. *Warships of the Royal Navy.* Annapolis: Naval Institute Press, 1979.

Morison, Samuel E., *The Two-Ocean War.* Boston: Little Brown and Co, 1963.

Palmer, Michael A. *Origins of the Maritime Strategy*. Annapolis: Naval Institute Press, 1988.

Polmar, Norman. *Atomic Submarines.* Princeton: D. Van Nostrand Co.. 1963.

— *Ships and Aircraft of the Us Fleet; Fourteenth Edition.* Annapolis: Naval Institute Press, 1987.

Polmar, Norman, and Thomas B. Allen. *Rickover; Controversy and Genius.* New York: Simon and Schuster, 1984,

Powers, Thomas. *Heisenberg's War.* New York: Alfred A. Knopf, 1993.

Preston, Anthony, and John Batchelor. *The Submarine 1578-1919.* Marshall Cavendish USA Ltd.

Price, Alfred. *Aircraft Versus Submarine.* London: William Kimber, 1973.

Rauft, Bryan, and Geoffery Till. *The Sea in Soviet Strategy.* Annapolis: Naval Institute Press, 1989.

Sapolsky, Harvey M. *Science and the Navy: the History of the Office of Naval Research.* Princeton: Princeton University Press, 1990.

Schratz, Paul R. *Submarine Commander.* Annapolis: Naval Institute Press, 1989.

Sharpe, Capt. Richard *Jane's Fighting Ships 1992-93.* Coulsdon: Jane's Information Group Ltd, 1992.

Shock, James R. *US Navy Pressure Airships 1915-1962.* Edgewater: Atlantis Productions, 1993.

Staff, Deputy Commander Submarine Force, US Atlantic Fleet *United States Ship Thresher (SSN-593): In Memoriam, April 10, 1963.* 1964.

Stern Robert C. *US Subs in Action.* Carrollton, Texas. Squadron/ Signal Publications, 1983.

Tarrant, V. E. *The U-Boat Offensive 1914-1945.* Annapolis: Naval Institute Press, 1989.

Treadwell, T. C. *Submarines With Wings.* London: Conway Maritime Press, 1985.

Tyler, Patrick. *Running Critical.* New York: Harper and Row, 1986.

Van der Vat, Dan. *The Atlantic Campaign.* Annapolis: Naval Institute Press, 1988.

Waters, Capt. John M. *Bloody Winter.* Annapolis: Naval Institute Press, 1984.

Watts, Anthony, ed. *Jane's Underwater Warfare Systems 1992-3.* Surrey: Jane's Information Group Ltd.. 1992.

Weir, Gary E. *Building American Submarines 1914-1940.* Washington: Naval Historical Center, 1991.

— *Forged in War; the Naval Industrial Complex and American*

Submarine Construction 1840-1951. Washington: Naval Historical Center, 1993.

Winslow, Richard E. III *Portsmouth- Built: Submarines of the Portsmouth Naval Shipyard.* Portsmouth, NH: Portsmouth: Marine Society, 1985.

Zubok, Vladislov, and Constantine Pleshakov. *Inside the Kremlin's Cold War; From Stalin to Khrushchev.* Cambridge: Harvard University Press, 1996.

INTERVIEWS

Interviews by Robert Largess, 1989 - 1998

Albacore CO's: Jon Boyes (with Harvey Horwitz), Bill St. Lawrence, Roy Springer, Lando Zech

Albacore Execs: Ted Davis, Lou Urbanczyk

Crewmen: Ron Polske (with Harvey Horwitz)

David Taylor: Jerry Feldman, Dick Stensen, Bill Day, Marvin Lasky, Sid Reed

CUW: John Coleman, George Wood

Stevens institute: Michael Bruno, Dan Sawitsky

Others: Anneta Stanton Kraus, Ned Beach (with Harvey Horwitz), Russ Van Billiard (with Harvey Horwitz), Charles B. Momsen, Jr., Harry Jackson (with Harvey Horwitz), Gene Allmendinger, Frank Andrews, Joseph Sawtelle, John Hart, Ray Hahn, Bob Johnston, Bruce Graves, Mark Anthony, Roy Syphers, Bob Fisher

Interviews by Harvey Horwitz, 1989 - 1993
Albacore Co: Ken Gummerson

Officers and Crewmen: Norman Bower, John McCarthy, Hines, Stan Zajechowski

Index

About the Authors

Robert P. Largess

A native of Washington D.C., Robert P. Largess attended Boston College and Brandeis University, where he earned an M.A. in Anthropology. He has taught in the Boston Public Schools - in some of the best and more of the worst schools in the city since 1968. At present, he is teaching English and World Literature at Boston Latin Academy.

He has published numerous articles and a short book on American social problems and urban education, appearing in-the *Washington Post*, the *Philadelphia Inquirer*, the *Boston Herald*, the *Pilot*, and the *American Educator*. He has also contributed to a curriculum guide on the history of the U.S. Labor movement, published by the Massachusetts AFL-CIO.

He has written numerous articles on naval and maritime history, appearing in *Naval Forces*, the *Submarine Review, Warship,* and Conway Maritime Press' annual volumes *Warship 1989, 1991, 1993,* and *1995.* These include histories of the *SS United States,* the *USS Triton*, and Lighter-Than-Air in the Navy, as well as the *Albacore.*

He has four boys and one girl, and lives in Boston with them and their lovely mother, Jeannine.

James L. Mandelblatt

Having become interested in history and submarines at an early age, Jim Mandelblatt has become deeply involved in the history of the submarine *Requin* (now on display in Pittsburgh, PA) and the men who sailed her.

He is the author of the book *Rebirth of a Submarine: A History of the U.S.S. Requin* (SS-481/SSR-481) and numerous articles on *Requin* as well as other submarines, and has been consulted for other articles on submarines. He is an active volunteer aboard the Liberty Ship SS *John W. Brown*, moored in Baltimore, Maryland.

Originally from Pittsburgh, Jim Mandelblatt now lives and works as a technical writer in the Washington, DC area. He is currently working on a re-issue of his first book on *Requin*, as well as in the concept development stage of books on life aboard *Requin* and on the radar picket and MIGRAINE programs.